HILARY GATTI is
the Università di Ro
books include *The R*
Knowledge: Giordano
The Natural Philosophy

Giordano Bruno and Renaissance Science

Giordano Bruno and Renaissance Science

HILARY GATTI

CORNELL UNIVERSITY PRESS

ITHACA AND LONDON

First published 1999 by Cornell University Press.

Printed in the United States of America.

Library of Congress Cataloging-in-Publication Data

Gatti, Hilary.
Giordano Bruno and Renaissance science / Hilary Gatti.
p. cm.
Includes bibliographical references and index.
ISBN 0-8014-3529-3 (hardcover : alk. paper)
1. Bruno, Giordano, 1548–1600. 2. Science, Renaissance.
I. Title.
Q143.B795G37 1998
509.4'09'031—dc21 98-27213

Cornell University Press strives to use environmentally responsible suppliers and materials to the fullest extent possible in the publishing of its books. Such materials include vegetable-based, low-VOC inks and acid-free papers that are recycled, totally chlorine-free, or partly composed of nonwood fibers.

Cloth printing 10 9 8 7 6 5 4 3 2 1

This book is dedicated

to the memory

of Mariano

Contents

Preface

This book reconsiders a dimension of Giordano Bruno's philosophy that has been ignored in recent years in favor of his Hermeticism and magic. From the beginning of my Bruno studies, however, it appeared to me that his attention was more often directed elsewhere, to subjects such as the new cosmology and the revival of ancient atomism, to number theory and the possibility of investigating, measuring, and mapping out anew the shape of the natural world. To stress these aspects of his thought could well seem a simple return to earlier readings of Bruno, which tended to treat with distaste his references to magic, astrology, and the art of memory, his cabalism, and his reading of the work of Raymond Lull. But was it really necessary to denigrate those aspects of Bruno's thought? I became convinced that Bruno's concern with such subjects, which has been so forcefully presented in the work of the late Dame Frances Yates, could be seen as surrounding and complementing his concern with the new science, particularly now that so many scientists of the early modern period, up to and including Newton, are known to have read extensively in such subjects and to have considered them integral parts of their culture.

Furthermore, was it necessary to see Bruno's sixteenth-century science as in all respects a prelude to the classical mechanical science that would dominate European culture from the early seventeenth century up to the end of the nineteenth? Might not some aspects of his thought seem closer to an age of post-Einsteinian relativity and quantum mechanics, with their theoretical justifications of scientific approximations rather than logical certainties and indisputable truths? Not until I had finished writing my book did I see Ramon G. Mendoza's *Acentric Labyrinth: Giordano Bruno's Prelude to Contemporary Cosmology* (1995), which is based on a similar insight into the remarkable modernity of some aspects of Bruno's scientific theory. Mendoza, however, devotes limited attention to the historical, renaissance context of Bruno's thought, which I believe must be the basis on which any claim for more modern intuitions is founded.

The greater part of this book was written in London during the academic year 1994–95, while I was on a year's study leave from the University of Rome "La Sapienza" in Italy. I am grateful to the University of Rome for granting me that period of leave and to University College London for hosting me during the academic year as an honorary fellow. At UCL I benefited from frequent and stimulating discussions on both Bruno and renaissance science with Giovanni Aquilecchia and Piyo Rattansi, to both of whom I express my gratitude. Arthur Miller's work on images and the imagination in scientific discovery, although not directly concerned with the renaissance, proved another source of stimulating thought.

Later on, the seminar "Corpuscular Theory from the Middle Ages to Newton," organized at St. Andrew's University by John Murdoch of the Department of History of Science of Harvard University and sponsored by the Foundation of Intellectual History, directed by Constance Blackwell, gave me an opportunity to consider Bruno's theory of matter in greater depth and to discuss related matters with specialist scholars. I am particularly grateful to Roger Ariew, whom I met on that occasion, for reading parts of my typescript and for advising me to approach Cornell University Press.

My attempt to restate the argument for a Bruno concerned primarily with the birth of modern science owes a particular debt to my daughter Anna, who in 1994–95 was making her own entry into the world of modern astrophysics. Her company and attention throughout my year in London make the writing of the book a pleasure to remember.

HILARY GATTI

Rome

Giordano Bruno and Renaissance Science

Introduction

The idea of a "scientific" Bruno, champion of the new cosmology and precursor of a Baconian investigation into natural causes outside the sphere of metaphysics and theology, has animated much Bruno criticism ever since the Nolan philosopher started, at the beginning of the seventeenth century, to be studied seriously by European scientists and philosophers.[1] It was an interpretation vigorously opposed by Frances Yates, who found her key to a reading of Bruno in occult philosophy and magic, particularly in the texts known as the Hermetica, which propound the spiritual vision of a divinized nature proposed by the mythical Hermes Trismegistus. It might have been possible to attempt an integration of a Hermetic Bruno into the scientific Bruno inherited from the nineteenth century. Occasional remarks by Yates suggest that at times she contemplated such a solution; for example, when she views the scientific revolution as taking place in two phases, first a magical-animistic phase, later a "mathematical universe operated by mechanics." The main thrust of her thesis, however, tended to deny this possibility, leading rather to a radical concept of incommensurability between the magical and the mechanistic worldviews: "The procedures with which the Magus attempted to operate have nothing to do with genuine science," she wrote in *Giordano Bruno and the Hermetic Tradition*.[2]

1. The standard bibliography of Bruno studies is Salvestrini and Firpo, eds., 1958. For a bibliography of more recent critical writings, see Ciliberto 1990. For the history of Bruno criticism from his death through the enlightenment, Ricci 1990.

2. Yates 1964.

Since Yates published her widely read and much admired book on Bruno in 1964, a great deal of work has led to substantial reappraisals both of Bruno and of the so-called scientific revolution of the sixteenth and seventeenth centuries. A lively reconsideration of Bruno has been developing, above all in Italy but with ramifications also in France, Germany, and Spain: a renewal of interest whose proportions have led it to be called a "Bruno renaissance."[3] Although these studies include various interpretations and diverse emphases, a pervading characteristic has been awareness of the need to reconsider Bruno's thought in the context of the larger European philosophical tradition (both as follower and as precursor of so much that has been central to that tradition) and not exclusively in the context of renaissance occultism and magic. An equally lively discussion has developed on the subject of renaissance science. This discussion was originally to a considerable extent stimulated by Yates's larger thesis that all renaissance science—and not only the philosophy of Bruno—was impregnated with and deeply influenced by Hermetic and occult texts and doctrines. The debate has expanded, however, to embrace the post-Kuhnian, skeptical philosophies of science and to meditate on the truth status that can be accorded to the new science in the present-day debate between "neo-Kantianism" and "neo-Hegelianism," with its radically relativistic historicism.[4] It is in the light of these revaluations both in Bruno studies and in the wider field of early modern science that this book reconsiders the relationship between them.

In Bruno's time, the word "science" was not yet common coin, and it would start to be used in the restricted sense we know today only by later figures such as Galileo. Bruno would still have thought of himself as a natural philosopher and of science as *scienza*, knowledge of any kind. Even so, such natural philosophers of the period as Paracelsus, Johannes Kepler, Tycho Brahe, William Gilbert, William Harvey, and later Galileo and Francis Bacon can be thought of as also practitioners of the new science, actively involved in more or less methodical research into natural causes and effects. Not everybody would agree that Bacon fits into that category. Certainly Bruno does not. He never made an astronomical observation of his own. His contribution to the rediscovery of ancient atomism was entirely theoretical, whereas only a few years later Kepler in Prague, Thomas Harriot in England, and Galileo in Italy would all be applying their understanding of the atomic theory to explanations of pre-

3. Ciliberto 1994:303.

4. Much of the substance of this debate can be found in Lindberg and Westman, eds., 1990.

cisely observed phenomena such as the refraction of light and magnetic attraction. Bruno contributed to such questions only through discussion and debate, although his contribution was often more technically refined than has generally been allowed. It also led to innovations of substantial and lasting importance in the theories concerned. Such innovations, however, were reached through a process of logical-philosophical reasoning, with a marked bias toward visualization through images and symbols, rather than through experiment or methodical observation. Bruno may most properly be described as a friend and supporter of the new science, or, in present-day terminology, as a philosopher of science, deeply concerned with its development and at the same time deeply suspicious of some of its outcomes and consequences.

Foremost among Bruno's suspicions of the new science was his well-known and clearly expressed distaste for the new mathematics, which he saw as a schematic abstraction attempting to imprison the vital vicissitudes of matter into static formulae of universal validity.[5] Copernicus, whom Bruno saw as a messenger of a new era to be lived in a universe no longer "out of joint," was in his opinion too much of the mathematician and too little of the physicist, unconcerned with an attempt to understand the patterns of force and energy in his new heliocentric world. Bruno, on the other hand, remained committed to Pythagorean number symbolism, which he linked to Euclidean geometry in an attempt to understand the transformations of matter in space. The result was a kind of mathematical physics in which number remained closely linked to the transformations of the material world. Ultimately, however, physics, biology, and the human sciences were fields of inquiry more congenial to Bruno than the abstractions of mathematics. It is perhaps no coincidence that he wrote a work of some scientific pretense, which can claim to be based on experience and observation, concerned with the restless and dynamic workings of the human mind. His brief and unfinished psychological treatise *De vinculis in genere* represents a remarkable attempt to demonstrate the ways in which language in all its forms acts on human behavior, influencing and even enslaving the will.[6] In this text Bruno's meditation on magic reaches its conclusion, concentrating on the powerful resonances that surround certain symbols, images, and words and on

5. Some comments on Bruno's mathematical thinking are to be found in Gatti 1989: 49–73. For Bruno's geometrical doctrine, a still valid study is Atanassievitch 1923/1972. Two major studies of Bruno's mathematics are in Aquilecchia 1993a. For recent studies of Bruno's mathematics related to his metaphysics, see Bönker-Vallon 1995 and 1997.

6. Published only in the nineteenth century in the third volume of the national edition of the *Opera latine conscripta*, 1879–91.

the way the mind reacts to messages, both verbal and symbolic, launched by those with the expertise and—only too often—the unscrupulous purpose of bending others' wills to their own. Bruno's awareness, at so early a stage, of the dangers inherent in the new forms of knowledge and communication which were gradually emerging in the sixteenth century may be seen as one of his most characteristic and original contributions to the development of the new science.

Another aspect of Bruno's meditation on the new science which is of great interest, both in the history of his own times and as a message for our own, is his awareness that revolutionary changes in the fundamental scientific theories that dominate a world picture, such as a new cosmology (in his times, the Copernican change from an earth-centered to a sun-centered universe) or a new theory of matter (in his times, the revival of ancient atomism), lead necessarily to a radical upheaval in the culture and society in which they occur. Not only was Bruno intensely aware of how the new science was "calling all in doubt." In the early pages of his Latin masterpiece, the so-called Frankfurt trilogy, published in 1591 but probably begun in England some years earlier, Bruno made a dogma of a systematic process of doubt to be applied to all traditional and orthodox ideas accepted by society:

> He who desires to philosophize must first of all doubt all things. He must not assume a position in a debate before he has listened to the various opinions, and considered and compared the reasons for and against. He must never judge or take up a position on the evidence of what he has heard, on the opinion of the majority, the age, merits, or prestige of the speaker concerned, but he must proceed according to the persuasion of an organic doctrine which adheres to real things, and to a truth that can be understood by the light of reason.[7]

If this introductory passage to Bruno's final and arguably greatest work, the Frankfurt trilogy, with its clearly pre-Cartesian tone, was admired and continually quoted by nineteenth-century commentators, to be virtually ignored by post-Yatesian commentators on Bruno's thought, that is no good reason not to take it seriously, even if today it has become clear that Bruno cannot be considered as simply a precursor of seventeenth-century rationalism.

Indeed, Bruno's thoughts on "truths that can be understood by the light

7. *De triplici minimo*, I, i, ibid., I, iii: 137–38; my translation.

of reason" were particularly complex. He was aware that the new science required the development of new processes of logic and new forms of language and communication: new words, he said, must accompany the new ideas about the natural world. Bruno, however, rejected the major new symbolic language that was emerging in his times, in the form of the new mathematics, and preferred to think in terms of a logic of pictures or images. Many commentators have seen this preference, clearly based on a well-known page of Aristotle's *De anima*, as linking him to a prescientific era. Frances Yates, in what was arguably her greatest work, *The Art of Memory*, was to lay much stress on this "prescientific" Bruno, whose art of memory she insisted on viewing, again almost exclusively, in terms of magical and occult mental resonances.[8] Rita Sturlese's more recent study of Bruno's major works on memory, *De umbris idearum* and *De imaginum signorum et idearum compositione*, however, has shown how the memory images in these texts actually work as systems of logical connection and communication—a discovery which suggests that Bruno's memory images should be considered as logical tools rather than as neo-Platonic or magical reflections in the mind of transcendental ideas.[9] Indeed, such a conclusion corresponds to the outcome of Bruno's meeting with the French king Henri III, to whom he dedicated the *De umbris idearum*. According to Bruno's account of this meeting at his trial, the king asked him if his memory techniques worked by magic or by art (scienza). Bruno claimed that he was able to demonstrate that they worked by art.[10] In today's computerized world, in which the visualization of knowledge through connected systems of images, icons, and symbols has ousted the now traditional mathematics as the unique and privileged language of science, Bruno's concept of a picture logic may seem less antiquated and more relevant to scientific inquiry than has usually been supposed.

Bruno's faith in the necessary progress represented by a methodical inquiry into the natural world was accompanied by a considerable dose of skepticism as to the possibility of the human mind's reaching exact and indubitable knowledge of the laws governing a universe defined in terms of an infinitely rich and varied process of metamorphosis and vicissi-

8. See the relevant chapters in Yates 1966.

9. See Sturlese 1990 and 1992b.

10. ". . . il re Henrico terzo mi fece chiamare un giorno, ricercandomi se la memoria che havevo et che professava era naturale o pur per arte magica; al qual diedi sodisfattione; et con quello che li dissi et feci provare a lui medesimo, conobbe che non era per arte magica ma per scientia": Firpo 1993:161–62.

tude.[11] Such skepticism would be out of step with the following centuries, which would be dominated by a more mathematical and mechanical concept of the universe and by an optimistic faith in scientific inquiry as a road to undoubted certainty and truth. Many commentators have deduced from this divergence between Bruno and early modern science that his thought about the natural universe was antiquated and irrelevant to the modern world. Ernan McMullin, for example, in an essay on Bruno's Copernicanism published in *Isis* in 1987, concluded that Bruno could be seen only as a late prophet of a neo-Platonic animistic universe in the last moment in which such a universe could be seriously proposed. But it may equally be possible to see Bruno as putting forward an idea of the universe and of the impossibility of the human mind's capability of reaching exact and final knowledge of its workings, which anticipates many ideas familiar to our own post-Einsteinian era, dominated by the approximations of relativity theory and quantum mechanics. One might see Bruno and Galileo as, in the final moments of that extraordinary cultural experience that we know as the Italian renaissance, bringing European culture to a crossroads. Galileo appears to us now undoubtedly as the pioneer scientist of the new era that was to pass through the idea of method and geometrical space of Descartes, Newton's mechanics, and on to the great practical successes of the eighteenth and nineteenth centuries. Bruno's proposals, on the other hand, for so long considered an abortive attempt to revitalize antique concepts of the art of memory and universal infinity, acquire a new relevance in the light of present-day post-Einsteinian debates on scientific theory and on the increasing uncertainty with respect to the kind of truths about our world which the scientist is able to achieve.[12] Some nineteenth-century commentators reevaluated Bruno as an early precursor of Darwinian evolutionary theory, and there seems to be no reason why other aspects of his natural philosophy should not be referred to more recent scientific debates. As Galilean and Newtonian science appear more and more to be related to a cycle of the scientific endeavor that closed at the end of the nineteenth century, Bruno's natural philosophy may seem increasingly to relate to scientific debates and concepts at the center of attention today.

Bruno pleaded constantly and coherently for the liberty of the new sci-

11. For interesting comments on these apparently contrasting attitudes in Bruno, see, for his historical optimism, Aquilecchia's introduction to the *Cena de le ceneri* (Bruno 1955), and for his epistemological pessimism, Badaloni 1991.

12. See on this subject Mendoza 1995.

entist to investigate the ways and laws of nature.[13] He was not disposed, however, as was Francis Bacon in his utopian *New Atlantis*, to see in those inquiries a necessary road to salvation. Bruno was intensely aware of the many ways in which the newly emerging sciences had become entangled with often conflicting plans for religious reform and moral regeneration. Some of his dialogues are wry satirical comments on that entanglement from the point of view of an unorthodox and non-Christian philosophical mind. His attempt to unravel these various strands of the intellectual life of his times led to a celebration of the new science, often in metaphors that would rapidly become standard in the literature of the time: the rising of a new dawn with the sun at last reestablished at the center; the new inquirer into nature figured as a latter-day Christopher Columbus voyaging into perilous, uncharted waters to discover a new world. But Bruno was already aware in the last decades of the sixteenth century that science would never be entirely free from religious and social attempts to condition its development. He also foresaw that the answers developed by the new science would not necessarily be final and true. On one hand, his early pro-Copernican dialogue, *The Ash Wednesday Supper*, contains one of the first appraisals of the new science (and specifically the new astronomy) as a path of necessary historical progress comforted by an ever-greater and more precise amount of carefully observed data. On the other hand, he repeatedly turns to a passage in the Old Testament book of Ecclesiastes which plays on the theme of the vanity of the human endeavor. This conscious ambiguity in front of the newly emerging inquiries into nature is clearly and beautifully expressed in the final image of his last and greatest cosmological work, the *De immenso et Innumerabilibus, seu de universo et Mundis* (1591). There Bruno claims that he has sojourned in the hills and valleys of nature, summoning forth their nymphs, symbols of the ordered forms of natural process. But the outlines of the nymphs have only gleamed briefly in the clear crystal air before fading back into their natural habitat once again. Such an image suggests that ultimately all scientific inquiry may be only a tautology, for it can do no more than attempt to describe what is there. In the end the mind is still faced by the enigmatic silence of the eternal hills and vales; for what is there is always so much less than what the mind needs to know.[14]

13. For Bruno's ideas on the liberty of thought and expression, from the time of his trial to the present day, see Gatti 1995a.

14. See the closing pages of the *De immenso* in Bruno 1879–91, I, ii.

From what has been said in the preceding pages, it should be clear that Bruno is to be considered one of the earliest philosophers of the new science rather than one of its practitioners. What characterizes most powerfully his treatment of the new disciplines is his concern with theory rather than with empirical observation. He reached all of his original contributions to the new cosmology, for example, on the basis of a theoretical premise that can be defined as an advanced and infinitely extended form of Copernicanism. This insistence on theory as a necessary premise of all scientific advance may have a familiar ring to the modern ear. It may also seem peculiarly out of step with a scientific process that in the seventeenth century would rapidly develop a mechanical world picture, insisting more and more on the value of empirical observation on the one hand and on mathematical proof on the other. It is no wonder that Bruno was pushed to the periphery of the new science until the advent of the new romantic philosophy, with its emphasis on mental paradigms accompanied by the concept of nature as a vitalistic process, which culminated eventually in the theory of evolution. Then, above all in the work of Friedrich von Schelling in Germany and of Samuel Taylor Coleridge in England, Bruno came back into the scientific picture as one of the propounders of a dynamic, vitalistic concept of matter.[15]

In our own century Frances Yates, in a widely accepted interpretation, read that very vitalism, with its consequent divinization of the natural process, as a Hermetic and mystical concept of the natural world. Paradoxically, this interpretation, based on precisely those elements of Bruno's thought which in the romantic era had readmitted him to the scientific discourse of the time, now had the effect of ousting him again from the field of the history of science. For in her evident distaste for modern empirical science Yates appeared to ignore the elements of romantic natural philosophy that had qualified Bruno as a distinguished forerunner and prophet, and her work was based on a desire to exempt him from any significant contribution to the field of modern science. Although her own contribution to Bruno comment and studies has been lasting and distinguished, investigating as it does little-known areas of Bruno's thought such as his Lullian texts and his art of memory and stimulating new and exciting areas of discussion, her appreciation of Bruno's relation to the new science has been largely negative and misleading. For Yates limited his contribution to the occult and Hermetic elements in his thought.

15. On Schelling's and Hegel's Bruno, see in particular Beierwalters 1973/1983 and Blumenberg 1975/1987; for Coleridge's Bruno, Gatti 1996a.

Bruno's technical capacity to participate in the scientific discussions of the period, stimulated by the observations of the new comets or by the reading of Copernicus's text, were, as I hope to demonstrate, constantly underestimated and misjudged.[16] Yates reduced even the theme of liberty of philosophical and scientific inquiry to little more than a question of moral integrity, unconnected to any intrinsic relationship to the nature of the scientific inquiry she supposed Bruno to have turned away from and derided.[17]

In this book I offer a reappraisal of Bruno's contribution to the scientific revolution that was already taking place in his time. I further relate some of his more scientific intuitions to the debate being developed by our own culture about the epistemological status of modern science. My primary claim is that the twentieth-century scientific discussion to which Bruno's work is most relevant is the conflict over theory—a conflict that is being worked out in precisely the field that Bruno thought of as his own, the philosophy of science. For both the more optimistic evaluation of theory as progress through refutation toward ever more refined and satisfactory premises (Popper) and the more pessimistic view of the necessary anarchy of method and the impossibility of exact observation of a chaotic world (Feyerabend) can find some comfort in Bruno's ambivalent meditations on the new science at the moment it emerged. It is both a historical Bruno immersed in the dramatic scientific developments of his day and a Bruno who has something to contribute to contemporary theoretical debate that this book will attempt to investigate and propose.

16. "Bruno is not at all in the line of the advance of mathematical and mechanical science. Rather he is a reactionary who would push the Copernican diagram or a compass invention back towards 'mathesis'": Yates 1964:324.

17. Ibid., 315.

PART I

Beyond the Renaissance Magus

1 "The Pythagorean School and Our Own": Bruno and the Philosopher from Samos

The epithet that Frances Yates gave to Bruno, the "Hermetic Magus," was not one that Bruno applied to himself. It is true that after the long quotation from the Hermetic *Asclepius* in his fourth Italian dialogue, *The Expulsion of the Triumphant Beast*, on which Yates bases much of her argument, Bruno's Isis calls the Egyptians, whose religion is being bewailed by Hermes, "my Egyptians." In the paragraph in which the earthly Sophia introduces this quotation from the discourse of Hermes Trismegistus to Asclepius, Bruno suggests why Hermes is lamenting the lost religion of the Egyptians, who "adored crocodiles, lizards, serpents, onions, as well as the earth, the moon, the sun and other stars in the sky, with magic and divine rites through which the divinity communicated most comfortably with men."[1] Yates interprets this passage as expressing Bruno's evident taste for a form of religion in which the divinity converses directly and eloquently with humankind through natural objects, even of the most humble and everyday sort. But one may question her claim that Bruno was predominantly and persistently influenced by Hermetic doctrine rather than by the other ancient schools of thought he so frequently recalls.

Bruno's Hermeticism, as Yates emphasized, has to be seen against the background of the renaissance doctrine of a *prisca theologia*: the yearning for the purity of unsullied origins which seems to have dominated the age

1. Bruno 1958:784. For an English translation of the *Asclepius*, which lies behind this whole passage of Bruno's *Spaccio*, see Copenhaver, ed., 1992:67–92.

in so many manifestations, from the neo-Platonism of the Florentine school to the Protestant Reformation. In an essential passage at the end of the first dialogue of his first philosophical work in Italian, written and published in London in 1584, *The Ash Wednesday Supper*, Bruno through his mouthpiece, Theophilus, theorizes this yearning in powerful terms. He is comparing those "like us," who are about to renew the old philosophy, with those like the pedantic Prudentius, who remain unquestioning disciples of the prevalent neo-Aristotelian, decadent, and warring Christianity. Let us judge, says Theophilus, on the basis of the results obtained by these two philosophies:

> The first lived in temperance, were expert in medicine, judicious in their thoughts, remarkable in their prophecies, miraculous in their magic, generous in their beliefs, observant of the laws, blameless in their morality, divine in their theology, in everything they did heroic; and this can be seen by the length of their lives, the strength of their bodies, their splendid inventions, the exactness of their prognostications, the substances transformed by their art, the peaceful cohabitation among peoples, their inviolable sacraments, the justice of their acts, their familiarity with good and protective intelligences, and the remains, which are still intact, of their extraordinary feats.[2]

The other sort, adds Theophilus, may be considered exactly the contrary.

Scholars of the renaissance such as Paul Oscar Kristeller, Eugenio Garin, D. P. Walker, Cesare Vasoli, and more recently Brian Copenhaver have shown how this vision of an ancient and pure school of thought, associated sometimes with the texts thought to have been written in most ancient times by a sage called Hermes Trismegistus and at others with semimythical figures such as Orpheus and Moses, runs through a long line of renaissance thinkers from Marsilio Ficino and Pico della Mirandola up to the later Symphorien Champier, Francesco Patrizi, and, in England, John Dee.[3] Bruno was taking over an already familiar theme. He was also bending and modifying it in important ways, however, as Frances Yates amply showed. For example, Bruno was far from sharing Ficino's interest in the *prisca theologia* as an anticipation of the future coming of Christ. Rather he expounded it as an alternative form of morality and religion. He was also considerably more radical and explicit than most of

2. Bruno 1958:44; my translation.

3. The discovery that the texts collected under the name of Hermes Trismegistus were probably written by diverse hands and belong to the early years of the Christian era rather than to ancient times, as renaissance readers believed, was made by Isaac Casaubon at the beginning of the seventeenth century. The story is told in Grafton 1983.

his contemporaries in using the eloquent praise of unsullied origins as a critical weapon to castigate the modern world.

A third consideration in respect to Bruno's adaptation of this traditional theme, however, escapes Yates's attention. In Ficino's preface to his translation of the Hermetic texts, which he called *A Book on the Power and Wisdom of God, Whose Title Is Pimander*, he wrote that Hermes Trismegistus "among philosophers first turned from physical and mathematical topics to contemplation of things divine, and he was the first to discuss with great wisdom the majesty of God, the order of the demons, and the transformations of souls."[4] It may well have been with Ficino's words in mind that Bruno referred to the Hermetic texts when, in *The Expulsion of the Triumphant Beast*, he was intent on defining a particular form of religious belief. Nevertheless, Bruno's own inquiry was by no means centered on theological questions. *The Ash Wednesday Supper*, like his later Latin masterpiece the *De immenso*, was a post-Copernican dialogue concerned with defining the structure and nature of the universe. When, immediately after his praise of ancient wisdom, Bruno mentions one of the most ancient philosophers as his particular source and model, it is not Hermes Trismegistus. Rather he turns to Pythagoras and terminates the first dialogue of his work with a discussion of great interest and importance on what he significantly calls "the school of Pythagoras and our own."[5]

It is clearly important that Theophilus's reference to his "school" of thought as neo-Pythagorean closely follows the celebrated pages of the first dialogue in which he praises Copernicus as the man who has liberated the world from a web of erroneous theories and "anchored contemplation to things that are constant and true."[6] Copernicus himself had emphasized the link between his own cosmological speculation and that of Pythagoras, declaring, in typical renaissance fashion, in his dedicatory letter of *De revolutionibus* to Pope Paul III, that rather than presenting a novelty he was restoring an ancient wisdom that had been lost.[7] So anxious is he that the pope should get his point that Copernicus quotes the page of pseudo-Plutarch which presents the differing cosmological doctrine of the Pythagorians Philolaus and Ecphantus.[8] Philolaus thought

4. Ficino 1975:50–51; quoted in Copenhaver and Schmitt 1992:147.

5. ". . . secondo l'uso de la scuola pitagorica e nostra": Bruno 1958:44.

6. Ibid., 28–29; Bruno 1977:86–87.

7. For the renaissance belief that truth was "old" rather than "new," see Rattansi 1996. For a detailed study of the Pythagorean "myth" among astronomers from Copernicus to Newton, see Casini 1994.

8. For a bibliography of Pythagorean sources, see Guthrie, comp., 1987.

that the earth, the sun, and the moon all orbited around a central fire in an oblique circle. Ecphantus also thought that the earth moves, but only around its own axis from west to east. Only when he had found such ancient authorities to support his ideas about the movements of the earth, protests Copernicus, did he start to consider such theories seriously.[9] The ten chapters of the first book of Copernicus's text, in which he presents a general description of his new universe, are crowded with further references to ancient doctrine, including the fleeting reference to Hermes Trismegistus which is made so much of by Frances Yates. This reference, however, is strictly limited to Copernicus's praise of the sun as center, which would rapidly grow into a cult of the sun, often with anti-Christian implications that certainly would not have pleased Copernicus.[10] It is of far less importance than Copernicus's reminder of Pythagorean precedent in the introductory letter to the pope. For Pythagoras, as modern commentators continue to underline, was the first to attempt the definition of a coherent cosmology. Moreover, he was a mathematician who thought of number as the fundamental principle of the world, linking its harmonies to the divinity itself.[11] Copernicus worked his reference to Pythagoras hard. In return he was often considered—sometimes with accusations of plagiary—as a disciple of Pythagoras, and his cosmology as a neo-Pythagorean universe.[12] It is significant that in 1576 Thomas Digges titled his pioneering English translation of the first book of Copernicus's text *A Perfit Description of the Caelestiall Orbes according to the most anciente doctrine of the Pythagoreans, lately revived by Copernicus.* Later, in 1615, when the Carmelite father Antonio Foscarini attempted (unsuccessfully) to uphold both Copernicus and Galileo by publishing a defense of the new cosmology which would succeed in reconciling it with the church, he titled it *The Pythagorean and Copernican Opinion of the Mobility of the Earth, and Stability of the Sun; and of the new System of Constitution of the World; in which he reconcileth the Texts of Sacred Scripture, and Assertions of Divines, commonly alleged against this opinion.*

9. See the dedication of the *De revolutionibus* to Pope Paul III in Copernicus 1543/1978.

10. For the renaissance cult of the sun, see "La rivoluzione copernicana e il mito solare" in Garin 1975.

11. There has been a recent upsurge of interest in Pythagoras as a serious philosopher and not just a symbol or a myth. See in particular O'Meara 1991, Gobry 1992, and, for the ethical dimension of Pythagoreanism, Thom 1995, an edition of *The Pythagorean Golden Verses,* which was frequently published in Latin throughout the sixteenth century.

12. For a specific and detailed study of Copernicus's Pythagoreanism, see Bilinski 1977.

The choice of Pythagoras, rather than Hermes Trismegistus, as the founder and model of his own "school" places quite a different emphasis on Bruno's cosmology. Hermes Trismegistus was thought by Ficino to have "turned away" from physical and mathematical topics to the contemplation of things divine. By founding her interpretation of Bruno's thought on his Hermeticism, Yates reached her celebrated conclusion that Bruno's Copernicanism should not be seen in the context of the history of astronomy but rather as "a hieroglyph, a Hermetic seal hiding potent divine mysteries."[13] This conclusion has been powerfully contested by Robert Westman, who claims that Bruno's cosmological dialogues, such as *The Ash Wednesday Supper* or the *De immenso*, are based on an attempt to develop a physical cosmology.[14] Westman's claim brings us back to the figure of Pythagoras, who, although his name was also associated with a form of mystery religion, cannot be said to have turned away from physical and mathematical topics but rather to have considered them essential to his doctrine.

The precise relationship between Pythagoras's mystical religion and the scientific aspects of his thought continues to be debated. In a still essential text titled *Lore and Science in Ancient Pythagoreanism*, Walter Burkert states that ancient Pythagoreanism recognizes him only as a religious "sage," although Burkert later admits that the lost book of Philolaus, the first Pythagorean to reveal the secrets of the sect in public, must be considered the authority on which Aristotle based his treatment of Pythagoreanism.[15] Carl Huffman, who treats Philolaus's thought in the context of pre-Socratic natural philosophy and cosmology rather than as an anticipation of Platonic trascendental ideas, tends to endorse the Aristotelian rather than the neo-Platonic interpretation of Pythagoreanism, in spite of the fact that Aristotle was critical of Pythagorean doctrine.[16] For Aristotle associates Pythagoras's doctrine with an infinite universe dominated by a warring struggle of contraries, and not with a transcendental sphere of Platonic ideas.[17] Moreover, Cicero remembered Pythagoras not as a mystic but rather as the first philosopher, indeed the inventor of the word itself.

13. See Yates 1964:241.

14. See Westman 1977.

15. See Burkert 1972, in particular chaps. 2 and 3.

16. See Huffman 1993.

17. Of Aristotle's frequent, although generally hostile, mentions of Pythagoras, see in particular *Metaphysics*, I, 985b23–987a28, in Aristotle 1984:1559–61. The Pythagorean doctrine of contraries, which Bruno was to take up and develop, is given a particularly detailed treatment in Teti 1988:82–83.

Pythagoras's traditional standing as the first of the philosophers cannot be overestimated in a consideration of the sources of his own inquiry, which Bruno chose to emphasize.[18] During the grueling questioning he underwent at the hands of the Inquisition in the long years of his trial, Bruno consistently defined himself as a philosopher. Such a definition formed an essential part of his defense. His philosophical inquiry formed the basis for his claim to a right to investigate questions traditionally considered the privileged territory of theologians: the nature of the soul in its relation to matter, the question of an afterlife, and the attempt to define divinity as first cause in relation to a universe which Bruno saw as infinite in extension and power. In an early but crucial session of the Venetian part of his trial, after a difficult debate on such questions, Bruno was asked if he had rethought these matters. The implication behind the inquisitor's question was clearly whether he had considered them in the terms required by Christian theological dogma and the Catholic Church. Bruno replied that yes, he had rethought such questions, but he saw no reason to modify his position; clearly, his philosophy had led him to certain conclusions which he had no intention of abandoning.[19] The long years of intricate wrangling which would follow, leading up to the final crisis and the sentence of death, were implicit in the trial from the beginning. For Bruno was claiming for the philosopher a principle of free thought and inquiry which implied an entirely new concept of authority: that of the individual intellect in its serious and continuing pursuit of an autonomous inquiry. When, during the last dramatic meeting with his judges, he told them that they might be pronouncing against him with more fear than he received his judgment, he knew that he was looking toward a future cycle of history in which no one church or system of dogmas would be able to dominate the life of the intellect, subjecting the individual thinker or philosopher to its will.

Frances Yates failed to understand the sense of Bruno's trial. The few pages dedicated to it in her book are among the least satisfactory and the most confused. It is to her credit that she realized this failing and attempted to justify her confusion about such an important issue. The trial

18. Cicero says in *Tusculanae Questiones*, V, 3, that Pythagoras should be considered the inventor of the term "philosophy." Dante takes up the reference in *Convivio*, III, 2, where he gives the etymology of the word (*filos* = *amatore, sofia* = *sapienza*) and writes: "per che notare si puote che non d'arroganza, ma d'umiltà è vocabolo. Da questo nasce il vocabolo del suo proprio atto, Filosofia." For the seven references to Pythagoras in Dante's works, see Toynbee 1894:376–84.

19. At the seventh hearing of the Venetian part of the trial, held on July 30, 1592. See Firpo 1993:196. For the relationship between Bruno's trial and a principle of free thought and expression, see Gatti 1995a.

remains fundamentally senseless, she claims, because the documents relating to Cardinal Bellarmine's crucial interrogation of Bruno on the eight heretical propositions to which his philosophy was ultimately reduced are missing. Furthermore, as was pointed out by Angelo Mercati, the theologian who introduced the official publication of the remaining parts of the Roman trial documents in 1942, philosophical and scientific propositions play little part in the interrogations we know about; the main body of the trial was occupied with theological queries and definitions of heresies. Mercati's analysis seems to have inspired scholars such as Antonio Corsano and Luigi Firpo to propose the possibility that Bruno was involved in a religious mission during the last years of his life, a thesis Yates embraces eagerly, even if the known documents fail to corroborate it.[20] This discussion confounds the real issue at stake—which may be considered as a definition of legitimate intellectual inquiry. Of course the inquisitors reduced to a minimum the philosophical and scientific questions on the agenda; for it was they who could impose the program, and their interests were primarily theological. There is some mention both of the new cosmology and of renaissance magic, for it was in relation to these subjects, among others, that Bruno's heresies were being defined. Bruno's trial was not "about" either of them, however, for it was not, as Yates correctly points out, like the later trial of Galileo, exclusively about the future of the new science. Nor was it about renaissance magic or a mystical religious mission. It was about free thought and the right of the philosopher to pursue an inquiry touching on the same subjects as those considered by the theologian. The church's position was that its subjects were its own territory over which it maintained absolute power and authority. Bruno's position was that the philosopher and the theologian could and should indulge in civil conversation while autonomously pursuing their differing intellectual inquiries, which, for the philosopher, could lead only where his individual reason directed.

It is impossible to understand the issue involved and to evaluate justly the stand made by Bruno with his life without appreciating the question of free thought and liberty of expression. His insistence on placing this issue at the center both of his work and of his defense is why Bruno remains so much a figure of the modern world. If there is, as many have argued, an intrinsic link between science and liberty of inquiry, then Bruno was among those who guaranteed the future of the newly emerging sciences, as well as claiming in wider terms a general principle of free thought and expression. It would nevertheless be an error to deduce that Bruno was a

20. Yates 1964:349–56.

democrat in the modern sense of the word. He was clearly also to a large extent a man of his times: an exponent of an aristocratic culture that had opened itself up to the idea of an aristocracy of merit, such as that contemplated by Sir Thomas More in his *Utopia*, but which nevertheless remained strictly anchored to the idea of culture as the domain of an intellectual elite.

The question of secrecy was fundamental to the *prisca theologia* both of the ancient world and of the renaissance. Yates makes much of the Hermetic cult of secrecy, which she sees as an essential aspect of Bruno's universe: that "Hermetic seal hiding potent divine mysteries of which he has penetrated the secret." Such a definition, once again, confuses the issue. It is true that the new science was deeply concerned with the problem of whether its ever more daring and dangerous discoveries should be communicated to the general public. Many of those concerned with the question thought not. Copernicus himself deliberately made his new cosmology difficult to understand, revealing it on his deathbed in a form so complicated that it was available to only a few. Bacon's House of Salamon in his utopian scientific society described in *The New Atlantis* reveals chosen discoveries only to the island community, keeping much of what it knows to itself. Galileo appears so modern undoubtedly in part because of his eagerness to use the printing press to communicate fully and clearly to the public the results of his scientific discoveries: an eagerness not appreciated by a church that still conducted a cultural policy of a very aristocratic and elitist flavor. But even where the new science adopted a policy of secrecy, often justifying itself through images of Hermetic secrecy and mystery cults, it did not become a branch of Hermetic doctrine, which, in the words of Ficino, turned away from physical and mathematical inquiry to investigate the workings of divinity and the immortal soul. It simply made use of a doctrine of secrecy available in the culture of the time.[21]

Bruno's major discussion of the problem comes, not by chance, at the very beginning of his series of philosophical works in Italian. It is significant that we find it at the end of the first dialogue of *The Ash Wednesday Supper*, in the pages immediately following Theophilus's definition of his school as Pythagorean. Clearly the secret doctrine referred to here is based on that of the Pythagoreans, whose secrecy cult involved strict rules and rites of which Bruno appears, to a large extent, to approve. He is aware, however, of the dangerous implications of a cult of secrecy and

21. Eamon (1994:5) remarks that "the Scientific Revolution exposed and neutralized nature's secrets."

makes the argument the subject of a lively discussion with his English counterpart Smitho, the cultured gentleman who represents in the dialogue the open-minded if still rather cautious intelligence of the English metropolitan middle classes which Bruno clearly hoped to capture as his ideal audience.[22]

This discussion consists of a long speech by Theophilus which stimulates an equally long objection by Smitho. Shorter exchanges then lead to the conclusion of the dialogue, where they agree to meet again the next day to continue their discussion of the Nolan philosophy. The characters of the two figures who conduct this discussion are relevant to the meaning of the passage as a whole. Theophilus is usually identified as an alter ego of Bruno himself, who conducts in his name the secondary narration and discussion of the Ash Wednesday supper. He claims to have been present at the supper, in the guise of one of the principal disciples of "the Nolan," as Bruno calls himself in this work. It is possible, however, that Theophilus does not always represent Bruno exactly. In this passage in particular he appears, in the first instance, to take up a particularly inflexible and extreme position on the subject of secrecy, which Bruno, by making it the subject of discussion, subjects to critical scrutiny. The critical voice in this case is that of Smitho, usually considered a mediocre sort of yes-man whose only role in the dialogue is to admire and agree with the Nolan philosophy as expounded by Theophilus.[23] Smitho's name marks him as an Englishman who is probably a typical rather than an extraordinary representative of the cultural world in which Bruno had been living for the past few months. Nevertheless, Bruno is careful to distinguish Smitho from the academic pedants from Oxford, who had interrupted his lectures at the university and who appear in the *Supper* as his principal Aristotelian-Ptolemaic opponents. Although Smitho seldom develops arguments sufficiently autonomous to attain a character of his own, at the end of the first dialogue he does appear as decidedly more than mediocre. His position is clearly distinguished from that of Theophilus when he expresses, at the crucial moment of the argument, a point of view that is even admitted as a possible correction of Theophilus (or Bruno) himself.

Having just praised the *prisca theologia* and condemned the moderns as their corrupt and decadent opposite, Theophilus adds that one sign of the decadence of the moderns is that they think they know everything.

22. The following pages develop a brief paper in Italian titled "Smitho: Un mediocre o un saggio?" read at the conference "Fonti e motivi dell'opera di Giordano Bruno" held on December 11–12, 1992, and published in *Nouvelles de la République des Lettres* 2 (1994): 197–200.

23. For this interpretation of Smitho, see Bruno 1955.

Smitho asks how such arrogance might be corrected. The first answer is put forward by Frullo, the impudent and lively servingman whose name in Italian ("frullare" = to whirl, spin, or whisk) indicates his delight in confusing solemn issues and turning philosophical arguments on their heads. Frulla's recipe for a cultural revolution is to cut off all the heads of the arrogant academics from Oxford, or wherever, and put new ones on their necks. Theophilus corrects him: it is not the heads that must be cut away but the mistaken ideas inside them. Here Theophilus recalls the Pythagorean school. What is needed is a good, solid course in philosophy in which all the schools of thought are presented together with the known scientific principles and discoveries. Like Pythagoras's students, nobody should be able to express a personal opinion before following the course to the end. Only when the foundations are solid can discussion lead to positive results.[24] Theophilus goes on to make a clear distinction between what he calls the Pythagorean method and that of the skeptic advocates of Pyrrhonism who make a profession of being unable to know anything. Their method is always and everywhere to ask and search for knowledge of something that they believe they will never find. Equally time-wasting and vain, in the opinion of Theophilus, is the activity of those pedants (represented in the *Supper* by the figure of Prudentius) who are always eager to dispute and cavil about the most trivial and marginal points. Only the Pythagorean school is valid because it is founded on a true desire to attain new knowledge of the universe supported by a rigorous intellectual discipline.

Theophilus's argument is powerfully and persuasively proposed and might have well been thought a good way of concluding the first day of the dialogue. Unexpectedly, however, at this very point Smitho claims that he has an objection to make. His perhaps overlong argument picks up an essential and telling point: in this process of teaching philosophical truth to new generations of pupils, who is going to be the master? Smitho admits to being perturbed by the relativity of knowledge. He quotes an objection familiar to skeptical thinkers from ancient times: has Theophilus not noticed how powerful custom is in inculcating belief? We are nurtured from childhood to believe certain ideas, and we remain attached to them so obstinately that they prevent us from seeing the most obvious truths. Our religion, our beliefs, our customs, even our science, are those of the society in which we are born. This being the case, will years of listening do anything else than form half-ignorant disciples of arrogant masters whose knowledge is only relative to the time and place in which they live?

24. For the pages under discussion here, see Bruno 1958:44–49; Bruno 1977:96–100.

Theophilus in reply takes up an idea which is exquisitely Pythagorean in origin. There are men sent by the gods, he assures Smitho, whose knowledge is pure and whose guidance is sure: an implicit reference to the Pythagorean sects who referred in hushed devotion to their master as "he who knows."[25] Such men, adds Theophilus, in what seems a curious contamination of ancient Pythagoreanism with modern Calvinism, are "elected" by the gods to illumine the inner spirit of their hearers with the best wisdom. But if Theophilus thought to make an English disciple by reelaborating in philosophical terms the Calvinistic doctrine of the elect, he is disappointed. Smitho finds such presumption highly dangerous, at least in the context of secular culture. Better, he opines, to go along with the opinion of the many, so that if there is an error it becomes a mistake of the society as a whole. And in a contorted passage, which has proved a stumbling block for translators, Smitho explains himself further. He believes with Theophilus that the truth is something precious which only a few possess. Sometimes, however, what the privileged few keep proudly for themselves is not true knowledge at all, but rather worthless prejudices that lead to folly and vice.[26]

Smitho is clearly making an important point. Privileged, secret knowledge can easily lead its owners into "follies de grandeur" that might destroy the world. Smitho's advocacy of making all knowledge general is thus not a cult of mediocracy but rather a safety valve for the community at large. Theophilus is far from despising Smitho for the objections he has made. *Bene*, he says, "well done." Still, in the future there will be some things which the multitude should perhaps better not know. In any case, continues Theophilus, the multitude has never discovered new knowledge for itself: this has always come from a few exceptional minds. Nevertheless, Theophilus concludes by conceding that this knowledge should be gradually and carefully released for the more educated of the public to understand.

Bruno's philosophical position at the end of the first dialogue of *The Ash Wednesday Supper* is strong. He has clearly identified the "school" that inspires his inquiry both at a general philosophical level and in his adoption and development of Copernicanism. For the Pythagorean cosmology not only supposed that the earth moved around a central fire but also

25. The *Life of Pythagoras* written by Iamblichus (c. A.D. 250–325) claims that stories about Pythagoras should be received as referring not to a mere man but to a superman. See Guthrie, comp., 1987:57–122.

26. "*Smitho*. Credo bene, che la verità è conosciuta da pochi, e le cose preggiate son possedute da pochissimi; ma mi confonde, che molte cose son poche, tra pochi, e forse appresso un solo, che non denno esser stimate, non vaglion nulla e possono esser maggior pazzie e vizii." See Bruno 1958:48.

extended the universe to infinite dimensions. By further situating the Pythagorean philosophy within a *prisca theologia* seen as the pure and unsullied origin of Western civilization, Bruno also associated his post-Copernican, infinite cosmology with a dimension of social order, peace, and justice based on a concept of harmony between the mind and the universe: "remember that your Pythagoras teaches that nothing in the natural world should be treated with disrespect," says the much maligned ass in Bruno's most satirical work, the *Cabala del cavallo pegaseo*.[27] Furthermore, beyond the individual mind tensed in the attempt to know the order and structure of the universe, Pythagoreanism proposed a first principle and cause in terms that were particularly congenial to Bruno, more congenial, he writes in his second Italian dialogue, *De la causa, principio e uno*, than the transcendental ideas of Platonic philosophy:

> Thus those who conceive of the principle of substance as one, conceive of the substances as numbers. Others, who conceive of the principle of substance as a point, conceive of the substances of things as geometrical figures. All of them agree in thinking in terms of a single principle, although Pythagoras's version is better and clearer than Plato's because his unity is the cause and reason of the individuality and the nature of points, and as an absolute principle it is more easily accommodated to universal being.[28]

It is interesting to see Bruno distinguishing so clearly between Pythagoreanism and Platonism; for renaissance Pythagoreanism was usually based on the neo-Platonizing accounts of Pythagoras's philosophy given in their *Lives* by Iamblichus and Porphyry.[29] Ficino and Pico della Mirandola both elaborate on this early neo-Platonic version of Pythagoreanism, as does Johannes Reuchlin, who considers Pythagoreanism as contemporary and similar to the the mysticism of Hebrew cabalistic doctrine.[30] Bruno seems here to be going back to more ancient sources that accentuated the nature of Pythagoras's first cause as less than transcendental: a kind of universal womb out of which the being of number and forms proceeds. Bruno's sources were probably Cicero's *Tusculanae Questiones*, the life by Diogenes Laertius, and above all Aristotle, whose treatment of Pythagoras, presumed to have been based on the lost book by Philolaus,

27. Ibid., 918.

28. Ibid., 331; my translation.

29. Porphyry (c. A.D. 233–305), the most brilliant pupil of Plotinus and the master of Iamblichus, himself wrote a brief life of Pythagoras. See Guthrie, comp., 1987:123–35.

30. Reuchlin 1517: "Quo animadvertimus Pythagorae philosophiam fere omnem esse a Cabalaeis ortam . . ." (fl. kiii r).

comes nearest to the version of his thought proposed by Bruno, even if Bruno is accepting as a positive philosophy a school of thought repudiated by Aristotle himself. The commentators of Pythagoras may continue to discuss among themselves, even today, what real justification there is for distinguishing Pythagoras's first cause from Plato's, and whether there is any evidence to suggest that Pythagoras was involved in scientific inquiry into the natural universe.[31] What interests us here, however, is that Bruno makes a clear distinction between Pythagoras and Plato, considering Pythagoras as the model and mentor for his own inquiry into nature. Pythagoras, he writes in another passage, was the first to search for "the traces and remains" of truth impressed onto the world of natural things. His first principle, rather than a sphere of transcendental ideas separated from nature, is a universal intellect whose ordered workings can be discovered at work within the natural world.[32]

Another aspect of Pythagoreanism which clearly appealed to Bruno was the way it allowed him, through the doctrine of metempsychosis, to maintain the immortality of the soul even after he had renounced his belief in the Christian religion and, more generally, in a fully transcendental God. There are times, particularly in the *Cabala del cavallo pegaseo* where the ridiculous ass remembers his previous life as the winged Pegasus, when Bruno treats this idea with humor and disrespect. At his trial, nevertheless, he candidly told his judges that he thought it could be the case.[33] Metempsychosis fitted in with Bruno's idea of matter and soul as the two interlinked dimensions of all bodies in the universe. Objects usually thought of as purely material were, in Bruno's opinion, endowed with minimal quantities of soul. Soul is thus present in all things, and, at their death and corruption, searches for new material forms in which to pursue the cycle of life.

Pythagoreanism contemplated an intermediary between God and the natural universe; but it saw that intermediary as the philosopher himself.[34] It is well known from the earliest accounts of the Pythagorean community that "the master" taught to most from behind a veil and revealed himself to only a limited number of chosen initiates. It is well known also that Pythagoras had traveled to Egypt and Babylon, participating

31. For a polemical argument against a scientific Pythagoras and in favor of the historical validity of the Egyptian sources of his thought, see Kingsley 1994.

32. Bruno 1958:1122.

33. ". . . il che se non è vero, par almeno verisimile l'opinione di Pittagora": Firpo 1993:176–77.

34. Iamblichus wrote that Pythagoras, instead of calling himself a wise man, called himself a "philosopher" (a lover of wisdom). See Guthrie et al., eds., 1987:66.

in the mystery religions that abounded in those cultures. Nevertheless, Pythagoras was revered by his disciples and followers as "the man" rather than as a god. Even so, Bruno dissociates himself from the most extreme implications of his ancient source in a page on which he also proposes for discussion one of the most crucial aspects of the new science: the question of the communication and revelation of its discoveries to the public at large. The rest of the work that follows constitutes Bruno's answer to that problem, proposing to the reading public a discussion of the Copernican cosmology in the form of a dialogue that calls into question virtually the entire society of the time: the Italian natural philosopher in the person of Theophilus-Bruno, the cautious but cultivated English middle-class intellectual in the person of Smitho, the pedantic and rigid neo-Aristotelian in the figure of Prudentius—recognized by Bruno as still very much a part of his world—and finally the lively if uncultivated Frullo, who represents the renaissance "common man."

Bruno's *Ash Wednesday Supper* narrates a revolution in the culture of the Europe of his time, caught in a moment of acute religious and intellectual crisis, by recalling that culture to its distant philosophical past and attempting to establish a meaningful continuity between its Pythagorean origins and the newly emerging science, particularly in its post-Copernican cosmological dimension. The doctrine of secrecy, however, with which Pythagoreanism was so closely associated, as the dimension in which the new science should most properly proceed, is repudiated, thereby revealing a paradoxical aspect of the Pythagorean experience itself. For if, on the one hand, Pythagoras's insistence on intellectual rigor and refinement led him and his followers to rebel successfully against the ignorant and arrogant princes of southern Italy and Greece, on the other the community's aristocratic elitism contains clear antidemocratic elements. It seems that Pythagoras's intense hatred of beans (which appear to have been excluded from the creatures of nature to be respected and revered) was based in part on the fact that they were used to count votes in elections.[35] Furthermore, Iamblichus tells us that Pythagoras was considered by his followers not as a mere man but as a superman. This is the starting point for Smitho's protest against a tendency which he rightly considers dangerous in the thinking about the new science and scientists of Theophilus-Bruno. His protest may be related to the fact that Smitho's own culture, although ruled by a powerful and domineering queen, was

35. Iamblichus notes these contradictory aspects of the Pythagorean philosophy. See ibid., 1958:98, 119.

beginning to develop a rudimentary form of parliamentary democracy. It is significant that Theophilus listens to, and at least partially accepts, Smitho's warning. Not all the "listeners" of Bruno's new course of natural philosophy will be among the few genuine "discoverers" of the new scientific society, but all of them must have the opportunity to know about the discoveries that are being made. For these reasons, Theophilus claims in the closing words of the first dialogue of *The Ash Wednesday Supper* that the Nolan philosophy will be strong enough to last. On its neo-Pythagorean foundations is to be built a liberal humanistic edifice that will enable it to resist the vain objections of the modern sophists and the blind prejudice of those anchored to the prevailing neo-Aristotelian religious and cultural dogmas.

The theme of traditional culture as an edifice which the *novatores* were trying to destroy to build a new one in its place was standard in this period. Montaigne, in his *Essays*, assumed a conservative stance: "to try to reshape such a vast bulk, to change the foundations of such a great building, is to imitate those who . . . wish to replace particular faults with universal confusion . . . and to heal sickness by producing death."[36] Descartes, in contrast, would argue that his new rationalistic philosophy would establish culture on entirely new foundations, a claim that has been questioned by John Cottingham, who argues that the new science failed to produce proposals that were as novel as Descartes presumed.[37] Bruno, by founding his new science on Pythagorean origins, is clearly trying to reconcile a return to the most ancient philosophy of an unsullied dawn with advancement toward new forms of knowledge of the natural world. He attempts to establish a continuity between the pre-Socratic science of the Greeks and a new inquiry into nature that finds its starting point in the Copernican revolution. The virulence and intensity of Bruno's anti-Aristotelianism has to be related to this historical vision. If Aristotle's master Plato was guilty of diverting attention from the world of nature to an improbable sphere of transcendental ideas, Aristotle himself was guilty of diverting attention to the overrefined abstractions of language and logic. These strategies, later crystallized by Ptolemy in the framework of a distorted cosmology, gave rise to what Bruno sees as long centuries of misguided error, rendered even more dramatic by the adoption of all these errors as the philosophical bases of Christian culture. Thus Pythagoras at one end of Western cultural history, and Copernicus

36. Montaigne, bk. III, chap. IX.
37. See Cottingham: 1993:145–66.

at the other, are the two true sources of light which Bruno celebrates in the first dialogue of *The Ash Wednesday Supper*. Illuminated by their beneficent rays, he will establish a radically restructured and enlarged, homogeneous and unified cosmological framework in which to found a new inquiry into both natural and divine truths. In this context, it is possible to see Bruno as a precursor of Newton, who also singled out Pythagoras as a key figure of the ancient past, not only as the founder of Greek science but also as the best preserver of the true religion.[38]

38. See Trompf 1991:230.

2 Discovering Copernicus

Any history of Bruno's discovery of Copernicanism should properly start with a careful consideration of book III, chapter 10, of his Latin masterpiece *De immenso et innumerabilibus, seu de universo et mundis*, in which the philosopher gives a brief account of his discovery of heliocentricity.[1] The *De immenso* is the third and last volume of the Latin trilogy of Bruno's final philosophical works, published in Frankfurt in 1591 and usually known as the Frankfurt trilogy. The earlier works in the trilogy are *De triplici minimo et mensura* and *De monade, numero et figura*. In the first of these works, which I will consider later, Bruno defines the terms of his revaluation of ancient atomism and his commitment to Euclidean geometry. In the second he discusses the concepts of traditional Pythagorean number symbolism, defining and commenting on the accepted meanings of the numbers from 1 to 10. It is clear, however, that his insistence on the importance of the monad, indicated in the title, links this work with the monadic or atomistic theories proposed in the first work of the trilogy. Only in the third and last work of the trilogy, the *De immenso*, is Bruno primarily concerned with cosmological speculation. In an important dedicatory letter of the whole trilogy to Duke Henry Julius of Brunswick, we can find a clear indication of the truth value which Bruno attached to the three works of which the trilogy is composed.[2]

1. Bruno 1879–91:I, i:389–98; in Italian translation, Bruno 1980:573. There is no English translation of this work.
2. For a comment on this letter in relation to contemporary politics and patronage, see Gatti 1989:20–21.

Bruno defines these works as follows: in the first (concerned with the minimum, the maximum, and measure), his doctrine and science arrive at a comprehension of the first elements of things; in the second (concerned with the monad, with number and form), revelation, faith, and divination trace the imprint of imaginings, opinions, and experience; in the third (concerned with the immense universe, innumerable worlds, and the infinite), certain and unequivocal demonstrations are offered concerning the disposition of the worlds, the unity of the infinite universe governed by a single principle, and the way in which, implicitly or explicitly, the natural order is revealed. In a brief concluding comment, Bruno emphasizes the order of certainty contained in these three works. The first expresses a desire to reach certain truth; the second searches for the truth not without uncertainties; the third attains truth without any shadow of doubt. These comments, whose vital significance to Bruno's philosophical inquiry has been too often ignored, make it clear that Bruno accords a very different truth status to each of the three aspects of his inquiry under consideration in the Frankfurt trilogy. The atomistic theory is presented as the most probable model for a theory of matter but is seen as desirable rather than certain: that is, it is presented as a viable hypothesis. The intuition of a metaphysical vision of unity within which to contain a theory of reality as numerical diversity is said to have been intensely experienced and argued, with both received opinion and past testimonies taken into account; but its truth status must remain uncertain and subject to debate. Only the cosmological propositions of the last dialogue are considered to have a satisfactory truth status as propositions that permit no shadow of doubt.

In the Frankfurt trilogy, Bruno inverted the order of his inquiry into the nature of the universe as it was developed in the first three Italian dialogues written in England between 1583 and 1585. There he began with cosmological speculation, based on the new theories of Copernicus: the starting point of his enquiry. Only in *Of the Cause, Principle and One*, and in *The Infinite Universe and Worlds*, which immediately follow *The Ash Wednesday Supper*, does Bruno develop his speculation in atomic terms. Even then, although an atomic theory is implicit in these works, and at times proposed explicitly, he does not yet dedicate a single work to a theory of matter in atomic terms. Rather than complementing his cosmological speculation with a suitable theory of matter, Bruno is concerned with fitting his Copernicanism into a new theory of the infinite in terms that will expand the physical universe to infinite dimensions, as well as defining what he calls "an infinite infinite": that is, a metaphysical principle of infinite order and unity which both transcends and defines, as cause and form of forms, an infinite universe of contraries and measure.[3] Bruno constantly refers to this infinite principle, with Pythagorean reminiscences, as his divinity, fountain of an order which is seen as the imprint on the world of nature of a perfectly ordered, divine mind.[4] From the years of the Italian dialogues, however, such a principle is clearly philosophical rather than theological, and its transcendence is dubious not only with respect to the Christian tradition but to renaissance neo-Platonism as well.[5] This aspect of Bruno's thought becomes more marked in the later Frankfurt trilogy, where atomism is seen as the foundation of the cosmological speculation. The search for the neo-Platonic unifying principle in the *De monade,* although developed no less intensely than in the Italian dialogues, is explictly admitted to contain what today would be considered a decidedly agnostic tinge.

To consider Bruno's cosmological speculation before his atomism and his search for the principle of unity is thus to disregard the order he himself gave to his philosophical speculation in its final expression. Such a disregard can be justified, however, both on the basis of the special truth status accorded by Bruno himself, in his final works, to the cosmological speculation over the other aspects of his inquiry and on the basis of chronological considerations. The Italian dialogues take the cosmological inquiry as their starting point. Furthermore, Bruno himself, in the tenth chapter of book III of the *De immenso,* offers a precise chronological account that shows how, in his youth, the cosmological inquiry was the starting point of the properly philosophical development of his thought.

Bruno's account of the development of his thought on cosmology is brief and worth quoting in full:

> It is well known that the earth does not change its inclination with respect to the pole, and by the ancients, among whom Timaeus and Pythagoras, it was held to orbit around the pole with an annual motion, while, with respect to the fixed stars, it was considered changeless and immobile (at least, generally speaking).
>
> Whoever considers the matter seriously will agree that those things which I imagined when I was young have been of little use except to illustrate the

3. For the infinite infinite, or "totalmente infinito," rather than a simple infinite quantity such as an infinite length or line, see Bruno 1958:382. The expression was later taken up by Thomas Harriot in his manuscript notes *De infinitis.* See Gatti 1989:61.

4. The influence of the neo-Platonism of Plotino here has been stressed by Beierwalters 1973; English trans. 1983.

5. On the vexed question of Bruno's pantheism, particularly as it was discussed by the German idealistic philosophers, see Blumenberg 1975/1987.

> relationship between the sun and the earth during the course of the year and to clarify the fact, which seemed important, according to which the moon, together with the earth completes an annual revolution around the sun. Then the desire took me to raise my inquiring gaze into a higher sphere, being certain that, given a slight shift of the earth from one pole toward the other, the position of our apse with respect to the apse of the fixed stars varied, giving rise to a new horizon. At that point I decided to consult Copernicus on this matter, having heard of him that he agreed with the ancients on the movement of the earth about the sun and about its own center, according to an annual and a diurnal motion. Then I realized that he had attributed motion also to the pole, just as I had imagined and learned from the ancients. According to Copernicus, the center of the earth moves between the poles of the zodiac, with a fourfold period.[6]

It is well to remember that this is an account of what Bruno considered to be, or wished to have considered, the early stages of his cosmological speculation as he saw it at the end of the intellectual process he is explaining. We cannot be certain that his Copernicanism actually developed in the terms proposed, but because it is the only account we possess of the stages of his reading of the Copernican revolution, it will have to serve as the basis of this history.

Bruno's account contains several interesting points. First, he claims that all the major aspects of the Copernican proposal were clear to him *before* he consulted the *De revolutionibus*. Whether or not this claim is true, it is based on a correct evaluation of the possibilities offered by the classical tradition from which Bruno claims to have received his heliocentric ideas. Plato's *Timaeus*, as Bruno had pointed out in *The Ash Wednesday Supper*, contains the germs of a theory of a heliocentric universe based on Pythagorean sources. Such sources would have been readily available to Bruno through later accounts such as Cicero's *Academica*, IV, 29, and Plutarch's *De placitis philosophorum*, III, 13, which he mentioned earlier in chapter 9 of this same book III of the *De immenso*: a chapter significantly titled "The Light of Nicolas Copernicus." There Bruno writes: "in Cicero I found that Nicetus had been aware of the movement of the earth and in Plutarch that also Ecphantus, Heraclitus, the Pythagoreans and Timaeus knew about it, so that on the basis of these sources I too began to consider the movement of the earth."[7] Through such classical sources, Bruno could have met, early in his intellectual de-

6. Bruno 1879–91, I, i:392–93; my translation.
7. Bruno 1879–91, I, i:384; Bruno 1980:566.

velopment, with both the idea of a diurnal revolution of the earth about its own axis and the annual orbit of the earth around a central fire or sun, as well as the possibility that the universe is to be considered infinite.[8]

At this point Bruno refers to certain "imaginings" of his own "when he was young": imaginings that should not be taken to mean mere chimera but rather attempts to visualize a new cosmological order. Bruno defines these "imaginings" as of little objective value and use, although he recognizes them as subjectively useful moments in his intellectual development. Through these visualizations, he claims, he attempted to clarify for himself the relations existing between the sun and the earth in the course of the year and also the relation of the moon to the earth in their annual revolution around the sun. In this page of the *De immenso* Bruno does not say what exactly these youthful "imaginings" were, although he had briefly outlined one of them in chapter 5 of book III of the *De immenso*.[9] Furthermore, an English text, which has been at the center of increasing critical attention, offers a surprising illumination with respect to this previously unglossed phase of Bruno's account.

William Gilbert of Colchester, author of the major renaissance text on magnetism, the *De magnete* published in 1600, the year of Bruno's death, died in 1603. Another work, *De mundo*, would only appear much later, in 1651, published by his half-brother, William Gilbert of Melford. The sixth book of Gilbert's *De magnete* attempts to integrate the author's study of magnetism into a coherent philosophical and cosmological vision of the universe. Stephen Pumphrey has claimed that Gilbert's work should not be seen as a scientific study of magnetism but rather as an attempt to develop a magnetic philosophy of the universe to oppose to the neo-Aristotelian philosophy of the schools.[10] Gilbert proposed his idea of magnetism as the cause of the movement of the planets within a somewhat cautiously Copernican or heliocentric image of the universe. In the *De magnete* there is little attempt to offer a precise cosmological picture of the new universe; and above all there is no explicit mention of Bruno.[11] In the *De mundo*, however, an interesting cosmological section offers in two or three pages, complete with diagrams, an account of two

8. For example, the astronomical fragments of Heraclitus of Pontus suggest that the earth rotates about its axis, from east to west, in one day; that the universe is infinite; and that the moon and stars are each a complete "world" floating in the infinite aether. See Gottschalk 1980:58–87.

9. Bruno 1879–91, I, i:360; Bruno 1980:546–47.

10. See Pumphrey's Ph.D thesis deposited at the Warburg Institute Library, London, for a detailed treatment of Gilbert's magnetic philosophy.

11. For Gilbert's cosmological doctrine in *De magnete*, see Freudenthal 1983.

kinds of motion of the earth put forward by Bruno. The first of these is presented as a speculation of the Nolan's *cum esset junior*, or "when he was younger."

Most previous mentions, in published works, of the two cosmological speculations ascribed by Gilbert to "the Nolan" have assumed that they refer to Bruno's first cosmological text, *The Ash Wednesday Supper*, a relatively youthful work compared to the *De immenso*.[12] A cursory look at the cosmological arguments developed by Bruno in the *Supper*, however, establishes that there is nothing there which corresponds to either of the models associated with the name of "the Nolan" in Gilbert's *De mundo*. Given the use of the phrase "when he was younger," it seems clear that the first of these pages of Gilbert's refers to one of the youthful "imaginings" mentioned, but not explicitly reconstructed, by Bruno himself in the passage of the *De immenso* quoted above on the terms of development of his cosmological inquiry. A return to chapter 5 of book III of that work bears out this hypothesis. For there Bruno explicitly reconstructs a cosmological model developed "when he was younger," which corresponds exactly to the one reproduced by Gilbert.[13] Given that we are dealing with a little-known comment on Bruno's cosmological speculation, it is worth giving this page of Gilbert's in full. The account is concerned only with the movement of the earth with respect to the sun and is developed around the diagram shown in Figure 1. The text, in English translation, runs:

> Let the earth move around its center on the axis of the equinoctial poles so as to create diurnal motion. Let the sun move with an annual motion around the earth along the equinoctial circle of the world. Let a third rectilinear motion be given to the center of the earth. For example: let C and D be the poles of the universe; E the center of the earth; AB the equinoctial circle of the universe along which the sun moves with an annual motion around the earth. However, let the center of the earth not always be where it is now, on the equator of the universe, but let it sometimes be in "e" closer to pole D and sometimes in "a" closer to pole C. It is evident that when the center of the earth is in E, it is a way of describing the equator of the earth when the sun is perpendicular; then again both tropics, namely the

12. The only full-length study of *De mundo* to date is Kelly 1965. For Gilbert and Bruno see in particular p. 39.

13. That Gilbert's reference to Bruno was in fact to the *De immenso* and not to *The Ash Wednesday Supper* had been correctly noticed by G. McColley in the first mention, in 1937, of these Brunian pages by Gilbert, and later on by Ricci 1990:54–55. See also Aquilecchia 1993a:324.

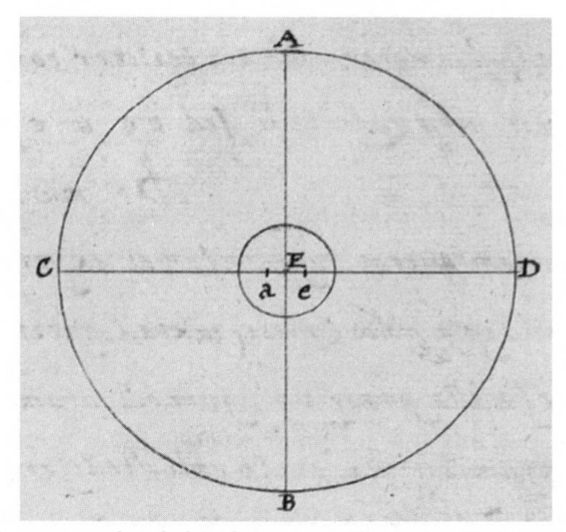

1. Diagram in the section headed "Alius movendi modus Nol: cum esset iunior" in William Gilbert, *De mundo nostro sublunari philosophia nova*. This diagram is in the manuscript copy presented to Prince Henry and held by the British Library, call mark: ROYAL 12 F XI, f. 87. By permission of the British Library.

> tropic of Capricorn when the center of the earth is in the northern part "e" and the tropic of Cancer when it is in the southern part "a." In between, however, there are intermediate parallel circles. But this rests on the assumption of rectilinear movement, which is not normally attributed to celestial bodies.[14]

This has all the appearance of a scholastic exercise, possibly used by Bruno in lessons on cosmology, delivered during his journey north from Naples. Its main point of interest is that it clearly constitutes a pre-Copernican phase of Bruno's thought on cosmology, or at least a phase that is not yet fully Copernican. Medieval speculators on cosmological questions had already suggested, also on the basis of ancient sources, the possibility that diurnal motion was caused by the revolution of the earth around its own axis; but because they still operated in the context of an Aristotelian physics, linked to the earth-centered Aristotelian-Ptolemaic cosmology, they had not gone so far as to put the earth itself in motion, but had it revolving at the center of the universe.

Bruno's proposition here seems to represent a very cautious move beyond this transitional hypothesis. Its principal characteristic is that it fails to make of the sun's orbit around the earth, known as the ecliptic, an oblique plane with respect to the celestial poles. Bruno's own account of

14. Gilbert 1651:199–201; my translation.

this youthful solution justifies this theory by claiming that, at that stage, he was concerned above all with facilitating calculation, although neither he nor Gilbert gives precise calculations to substantiate this claim. To ensure the movement of the sun between the two tropics, Bruno is thus obliged to make the earth move to the left and right of its equatorial position, a solution that would immediately raise the objection that the movements of celestial bodies were supposed always to be circular and never linear. Gilbert's account does in fact raise this traditional neo-Platonic objection. Bruno's account continues by claiming that the uniformity and perfect circularity of celestial motions is pure fiction: a thesis which Kepler would later prove, somewhat to his own horror and dismay. Bruno goes on to state what was clearly a youthful and never completely abandoned conviction that the attempt to reduce celestial motions to exact calculation was doomed to failure. The full implications of this conviction will be discussed in another section of this book. What should be noticed at this point is that Bruno's early geocentric cosmological proposal contains two important elements. First, it contemplates a minimum movement of the earth from its central position in the cosmos. Second, it visualizes such a movement as linear, underlining the fictional status of the regular circular movements of heavenly bodies still accepted by Copernicus. Furthermore, Bruno's slightly sad recognition in the *De immenso* that his youthful cosmological speculation went almost unnoticed turns out to have been unduly pessimistic. It clearly interested Gilbert enough to wish to reproduce it. Once the stages of Bruno's own cosmological inquiry have been clarified, it will be worthwhile investigating briefly the terms of Gilbert's cosmological speculation and attempting to define the reasons why he gave particular value to Bruno's contribution.

According to the passage in the *De immenso* quoted above, Bruno claims to have developed his initial variations on the order of the earth-sun relationship entirely on the basis of ancient sources. These offered him cosmological solutions ranging from the Aristotelian-Ptolemaic geocentricity, accepted throughout the Christian Middle Ages, to various forms of heliocentricity. Bruno's claim to have moved from a geocentric cosmology to a heliocentric one independently of Copernicus's book is surprising but not impossible. Furthermore, Bruno's decision to consult Copernicus's book appears quite clearly not to have arisen through any doubts concerning the heliocentric hypothesis as such, but rather through doubts about a further and most crucial question within the astronomical speculation of his time: the so-called precession of the equinoxes.

The precession of the equinoxes is the name given to the apparent slow shift in the equinoctial points, which are the positions in the heavens that are directly overhead on the two occasions of the year when day and night have equal lengths. This happens in the spring and autumn, giving the spring or vernal equinox and the autumn equinox. The precession of the equinoxes has to be understood with respect to the phenomenon stated by Bruno at the opening of his account: that is, that the oblique axis of the earth (in a heliocentric cosmology) remains stationary in its obliquity throughout the year with respect to the celestial poles. If the earth's axis were truly fixed in orientation, relative to a permanent starry background, the location of the vernal and autumn equinoxes, measured against the starry reference frame, should be unchanging. That is, at the moments when the day and night are of the same length in spring and autumn, the positions of the starry framework seen from earth should remain unchanged every year. In the second century B.C., the Greek astronomer Hipparchus discovered that this is not in fact the case and that the equinoxes regress very slowly through the zodiac. The rate of precession is so slow that it is impossible to observe with the naked eye in the course of a lifetime. Hipparchus discovered the phenomenon through measurements taken at an eclipse of the moon, which he compared to those taken by his predecessor Timocharis during a previous lunar eclipse. What interests us here is that Bruno knew the work of Hipparchus, which he explicitly mentions more than once in the *De immenso*. This suggests that Bruno may well be giving an accurate account of his development as a cosmologer when he claims that he first learned of the precession of the equinoxes, as he learned about heliocentricity, through ancient sources. Nevertheless, it was when he decided to "raise his inquiring gaze" into these higher regions of the zodiac that he found he needed guidance which he had not required before.

Bruno was not an observational astronomer. The few measurements reported in his astronomical writings were taken from others. He was also deficient in mathematical training, although he appears to have understood, somewhat reluctantly, the necessity of supplying a mathematical account of the celestial phenomena. Hipparchus's discovery of the precession of the equinoxes had already been absorbed by Ptolemy in his earth-centered post-Aristotelian astronomy developed in the *Almagest*. Bruno's problem seems to have been that of understanding how the phenomenon of the precession of the equinoxes could be satisfactorily resolved in terms of a heliocentric universe. According to his account, this technical problem led him to consult Copernicus's book, and not the much larger problem of heliocentricity itself, which he claims to have already understood from the ancients. He was not disappointed. He found that Copernicus had taken ample account of the precession of the equinoxes, absorbing the slow movement of the equinoctial points into a heliocentric account of the celestial phenomena.[15] The Copernican astronomy explains the apparent slow slipping back of the vernal equinox by assuming that the orientation of the earth's axis in its revolutions around the sun is not constant but is very gradually altering in a periodic fashion, each complete cycle taking approximately twenty-six thousand years. Copernicus was the first to supply a reasonably accurate value for the rate of precession, although he erroneously believed this rate to be variable with time. Bruno clearly studied Copernicus's account of precession with particular care, for his own accounts of this complex phenomenon when he describes the various movements of the earth in his infinite universe are correct and, on the whole, precise.

We have no precise data to help us determine when and where Bruno discovered Copernicus's *De revolutionibus*. Frances Yates, in her book of 1947, *The French Academies in the Sixteenth Century*, said this discovery took place during his years in Paris between 1579 and 1582. She leaves no doubt as to the deep interest in the new cosmological speculation in the circle of academicians and poets at the French court. Bruno was officially employed in Paris by the French king Henri III to give lessons as a *lecteur royale* and so would have been close to the court culture of the period.[16] Yet Gilbert's mention of Bruno's youth when he started speculation on cosmology, added to the scholastic nature of the models Gilbert supplies, suggest that he may already have been conversant with the subject before he left Italy in 1578 at the age of thirty. This supposition is supported by Bruno's claim to have taught the *Sphaera* (almost certainly a reference to the Aristotelian-Ptolemaic *Sphaera* of his fellow Dominican Giovanni di Sacrobosco, which remained one of the most popular astronomical textbooks throughout the sixteenth century) to pupils in the region of Genoa on his journey north through Europe.[17] Recently a copy of Copernicus's *De revolutionibus* was found in the Biblioteca Casanatense in Rome with the inscription "Brunus" on the flyleaf. The handwriting bears no resemblance to the few surviving samples of Bruno's normal hand. This does not seem particularly relevant to an identification in this

15. See Swerdlow 1975:49–98.

16. Following and developing Yates 1947, Gosselin 1988:166–81 has underlined the importance of Bruno's French experience.

17. For Bruno and the *Sphaera* of Sacrobosco, see Canone, ed., 1992:79–83.

case, however, as the inscription is highly stylized and ornamental, in the way students inscribe copies of their favorite textbooks. Within the circling wreathlike design that surrounds the name "Brunus," one can distinguish the letters FrD, which could indicate that the book was acquired by Bruno after he entered the monastery of San Domenico in Naples in 1565. Later on in the text, a different hand has made what seem to be official expurgations and corrections. These are dated after Bruno's death and were clearly made in an attempt to render Copernicus's text, which would be put on the Index of forbidden books in 1616, theologically respectable to read. Although the precise ownership of this book remains uncertain, it appears to have been part of an ecclesiastical collection. This suggests that it could have been among the books confiscated when Bruno began to get into trouble at his monastery.[18] Bruno might first have read Copernicus in a very youthful period indeed, perhaps shortly after he started studying at the Dominican monastery in Naples in 1565. In that case, he would have been well aware of Copernicus's cosmology before he arrived in France and been prepared to absorb the influences circulating in Paris after the translation of the first book of *De revolutionibus* into French by Pontus de Tyard under the title *Deux discours des parties du monde.*[19]

The works published by Bruno in France, however, are not primarily concerned with cosmological questions. His first published work to have survived, the *De umbris idearum* of 1582, does place the sun at the center of the universe. However, possibly influenced by Tyard, it establishes the point in mystical and Hermetic terms as part of a renaissance cult of the sun, rather than supporting it with specific cosmological considerations. It is only with Bruno's arrival in England that we find a properly cosmological speculation moving to the center of his inquiry.

18. The first mention of this volume is to be found in McMullin 1987:59. Both McMullin and Granada 1990:358–59 assume the volume to be Bruno's, although the attribution is not certain. The frontispiece, with the name "Brunus" and the accompanying signet marks, has been reproduced in Canone, ed., 1992:93.

19. Pontus de Tyard, bishop of Chalons, published his book in 1578. See Yates 1964: 174–75.

PART II

Toward a New Science

3 Reading Copernicus: *The Ash Wednesday Supper*

It remains a debated point whether, in the disastrous series of lectures which Bruno attempted to give at the University of Oxford shortly after his arrival in England in 1583, he proposed as his central theme the Copernican astronomy. The documentary evidence offers only one extended account of the dramatic episode, which saw Bruno's lectures interrupted by dons ostensibly scandalized because Bruno had plagiarized the works of Marsilio Ficino. The account by George Abbott, a fellow of Balliol College at the time of Bruno's lectures, and later archbishop of Canterbury, suffers from the double disadvantage of being a much later evocation of the episode (published in 1604) and of being a highly loaded and virulent piece of anti-Italian and anti-Catholic polemic.[1] Abbott starts his account by claiming in disapproving terms that Bruno was talking about Copernicus, attempting to "set on foote" the opinion that "the earth did goe round, and the heavens did stand still." This was the first attempt to teach the still theologically condemned Copernican astronomy at Oxford, apart from a brief reference in some lectures given by the distinguished mathematician Henry Savile in 1573. As Giovanni Aquilecchia has demonstrated, however, Savile introduced Copernicus only inci-

1. Abbott's page on the Oxford lectures of Bruno is to be found in his work of religious polemic titled *The Reasons Which Doctour Hill Hath Brought, for the Upholding of Papistry, Which Is Falselie Termed the Catholic Religion* (1604:88–89). Discovered by Robert McNulty in 1960, it is quoted in full in Yates 1964:208–9. For an account of this page on Bruno in the context of Abbott's work as a whole, see Aquilecchia 1993a:246, and for a more fully documented discussion of Bruno's Oxford experience, Aquilecchia 1993b and 1997.

dentally in a calculatory context, without suggesting a realist interpretation of his astronomy.[2] That Bruno in 1583 should, possibly for the first time, have attempted a realist reading of Copernicus in such a renowned but conservative public place clearly scandalized Abbott, who accuses him of being quite mad. Nevertheless, since the discovery of Abbott's page by Robert McNulty in 1960, Bruno commentators have tended to privilege Abbott's reference, in the second part of his account, to Marsilio Ficino, whose *De vita coelitus comparanda* Bruno was, according to Abbott, accused of plagiarizing, giving rise to the interruption of his lecture course.[3] Although Abbott himself offers no indication of the terms in which Bruno was reconciling his Copernican *exploit* to the pre-Copernican neo-Platonism of Ficino, Frances Yates emphasized this latter part of Abbott's account as proof of her Hermetico-magical thesis, pointing out that the Ficinian text in question is concerned with "astral magic and sun worship" and not with astronomy or cosmology.[4] Rita Sturlese has commented on the Oxford episode in the context of a complex thesis that proposes Ficino's neo-Platonic philosophy as it had just been expounded by Bruno in a Latin work, the *Sigillus sigillorum* published soon after his arrival in England, as the foundation on which, although in disagreement with Ficino's transcendentalism, Bruno constructs his own philosophical edifice, prefigured in his Oxford lectures titled (as he informs his readers in Dialogue IV of *The Ash Wednesday Supper*) *de quintuplici sphera* and *de immortalitate animae*.[5] The major problem presented by both these explanations of Bruno's Oxford experience is that Abbott starts off his account with a critical comment on Bruno's Copernicanism and not on his magic or his neo-Platonism. Furthermore, Bruno himself specifies as its historical and intellectual sequel a "supper," said to have been hosted by Sir Fulke Greville in London after Bruno's return from Oxford, in which he was invited, not to preach Hermetic magic or mysticism, or to discourse about Ficinian neo-Platonism, but rather to "explain his Copernicanism and other paradoxes

2. See "Copernico nelle lezioni oxoniensi di Henry Savile (1573 ca.)" in Aquilecchia 1993b:376–82. Aquilecchia is replying to a claim made by Mordechai Feingold in Vickers, ed., 1984:87 that Savile's Oxford lectures included "a long and detailed account of the Copernican theory."

3. Reported in *Renaissance News*, 13 (1960):300–305.

4. Yates 1964:209.

5. See, for this thesis, Sturlese 1994.

6. The somewhat contorted translation of this passage in Bruno 1977:85 ("he [the retainer] longed for his conversation on [and thus his exposition of] Copernicus and other paradoxes in his new philosophy") appears to depend on an attempt to integrate the final

of his new philosophy."[6] Fulke Greville supplied Bruno with two specially chosen neo-Aristotelian and anti-Copernican opponents, "presumed" by Bruno to be from Oxford; the bitter references to his Oxford disgrace in the text clearly suggest a close link between the two events.

In the attempt to solve what Aquilecchia considers the "equivocal point" raised by Abbot—or the unspecified connection between Bruno's Copernican discourse at Oxford and the pre-Copernican and largely magical text of Ficino's which Abbot specifically mentions—help can be gleaned from the literature on magnetism which I shall be considering in more detail, with reference to Bruno, in Chapter 6. Here it is enough to refer to a text published in 1629, in Ferrara, by the Jesuit Niccolò Cabeo, *Philosophia Magnetica*, remembering that Cabeo was considered by Gad Freudenthal as the magnetic philosopher William Gilbert's "most thorough reader (and critic)."[7] In pages of decisive repudiation of the post-Copernican cosmological implications which Gilbert, largely in the wake of Bruno, had drawn from his magnetic philosophy of matter, Cabeo refers specifically to the *De vita coelitus comparanda* of Ficino as a text that influenced Gilbert by extending the magnetism concentrated in the poles of earth to the polestar itself, and through it to the constellation of the Little Bear. Cabeo, who does not mention Bruno, was clearly thinking of chapter 15 of Ficino's text, where, writing of the nautical compass, Ficino remarks that the lodestone draws the needle toward the Bear, "since the power of the Bear prevails in this stone; and from there it is transferred into the iron and pulls both toward the Bear."[8] This extension of an earthly, material phenomenon such as magnetism to the stars above the planetary system already tended to deny the Aristotelian-Ptolemaic distinction between an unchanging heavenly fifth essence and sublunary elemental matter; and Cabeo is every bit as indignant about such a denial as Bruno's Oxford critics seem to have been nearly half a century earlier.[9] Ficino, for his part, had clearly been thinking about magnetism as a primarily magical, occult phenomenon. Nevertheless, he appears to have been aware of the dangerous implications of his magnetic hypothesis, as in a later *Apologia* or defense of this text, dated Careggi, September 15,

version of the *Supper* with a primitive version which says that the royal retainer "desired his [that is, the Nolan's] conversation, particularly because of the desire he had to understand the reasons for the movement of the earth, and other paradoxes which he [the Nolan] fully approved of." For the original of these two versions of this passage of the *Supper*, in the text recently established by Giovanni Aquilecchia; see Bruno 1994b:37, 299.

7. See Freudenthal 1983:37.

8. Ficino 1989:315.

9. Cabeo 1629:25–28.

1489, he anticipated not a few "malicious" criticisms: "And someone . . . , unworthy of life, will begrudge life to the heavens."[10] Although there must remain an element of speculation about whether Bruno actually linked Ficino to Copernicus in these magnetic terms at Oxford, there can be no doubt that his first philosophical work to be written in his native Italian was substantially a dialogue about the cosmological implications of Copernicanism. Bruno himself mentions it to his judges, during the Venetian part of his trial, defining it as a work "divided into five dialogues which treat of the movements of the earth."[11]

Bruno's *Ash Wednesday Supper* is not an easy book to read. It contains his first attempt at a methodical exposition of his understanding of Copernicus's cosmology, which is immediately defined as a critical understanding. From the beginning of his text, Bruno makes quite clear his intention of going beyond Copernicus on several points that would turn out to be at the center of much of the astronomical and cosmological discussion in the following century. Bruno's major innovation is his refusal to accept our own solar system, with its then five known planets besides the sun, moon, and earth, encircled by a sphere of stars, as representing the full extension of universal being. This daring innovation went well beyond what was explicitly suggested by Copernicus's book. For although Copernicus may have been aware of the destructive implications of his new cosmology with respect to the Aristotelian-Ptolemaic cosmology and physics, he never spelled out those implications himself. Rather, he deliberately appeared to accept the basic shape and tenets of Aristotle's universe as proposed in his *De caelo*, closed within a system of revolving planetary spheres surrounded by a sphere of fixed stars. Copernicus gave the appearance wherever possible that he was simply reversing the relative positions of the sun and earth.[12] Bruno pushes the realization that the new heliocentric cosmology actually opens up the universe to far greater dimensions than the previous Aristotelian-Ptolemaic cosmology had contemplated to its extreme conclusion. He proposes an infinite universe populated by an infinite number of worlds: a proposal already outlined in the *Supper*, although it would be fully developed only in the following

10. Ficino 1989:395.

11. Firpo 1993:188.

12. The deliberately cautious rhetoric that Copernicus employed in his effort to present his new astronomy in a way that would not seem revolutionary or lacerating to his fellow churchmen and other contemporaries is underlined in Westman 1990:167–205. Granada 1990 has persuasively argued that Bruno's own Copernicanism owed much to the *Narratio prima* of the Protestant and more polemical Rheticus.

two dialogues, *The Cause, Principle and One* and *The Infinite Universe and Worlds*.

Bruno's infinitism was, perhaps, his major contribution to the cosmological discussion of his time, and it has remained at the center of critical attention from his day to ours. It is intimately connected to the fraught question of the planetary orbs, which Copernicus had not explicitly denied but which would be gradually dismantled at the end of the sixteenth century in a process that has been much discussed in several recent texts.[13] The planetary orbs may be defined as the space bounded by two concentric circles within which each planet carries out its specific revolutions around the center. Originally conceived of as aerie and subtle, the orbs gradually came to be defined as composed of some solid substance onto which the planets were in some sense fixed and carried round. Bruno in the *Supper* expresses very clearly his profound disbelief in the "obscure" and "monstrous" solid orbs, whose existence he rightly claims is not to be found in Aristotle himself. Later, in the *De immenso*, he repeats this point, claiming that the solidity of the orbs was first suggested by certain "astrologers" of antiquity, and in particular by Averroës.[14] Furthermore, in the *Supper* Bruno begins to reflect on a possible movement, or movements, of the sun at the center: a theme to which he will return in the later *De immenso*, and which became increasingly important in the seventeenth-century astronomical discussion after Galileo's sighting of sunspots.

The difficulties of *The Ash Wednesday Supper*, however, do not derive simply from the fact that the book contains an homage to and original reading of Copernicus, together with a critical evaluation of some of the basic Copernican assumptions. They derive also, and perhaps principally, from the complicated literary and conceptual context in which Bruno develops his cosmological speculation. This is not the place to comment on the structural and literary quality of this remarkable text; although any full appreciation of its multiple meanings would have to take such factors into account. It must, however, be emphasized here that another dimension of Bruno's first Italian dialogue is the meditation on the nature of the new scientific truths and their particular status as truths, which is intertwined with the cosmological speculation. What we have in the *Supper*

13. See in particular E. Grant 1994; Goldstein and Barker 1995.

14. Sturlese 1992a has commented at length on Bruno's reading of Averroës, but she fails to pick up the particular point about the orbs. For Bruno's remark about Averroës in this context, see *De immenso*, bk. III, chap. 7, in Bruno 1879–91, I. i:370.

may be seen as both more and less than the relatively straightforward exposition of Bruno's cosmology in the later *De immenso*. The *Supper* is a dialogue in the renaissance tradition in which argument and polemic are used in the most radical terms to establish a deliberately heated debate.[15] The debate is heated because it involves much more than a question of alternative cosmological models. It involves a physics, and beyond that a metaphysics, which had become established over the centuries as cornerstones of cultural, theological, and even political power throughout Europe. To upset it was to upset a world, with its accepted language, certainties, and assumptions. It was also, as Bruno was acutely aware, to propose a new science whose truth value and paradigms had still to be discussed and established. The part of his argument concerned with this subject will also have to be deferred to another section of this book. For the moment it is sufficient to point out that *The Ash Wednesday Supper* is an anomalous and complicated text. Its contribution to the new science is manifold and has been underestimated. It is to a large extent, however, a dialogue *about* the new science and about those who were attempting to suffocate it, rather than a text *of* the new science presenting methodical and reasoned conclusions. Its cosmological argument has to be extracted from a richly but at times excessively complex literary and rhetorical background. It is thus not altogether surprising that the Copernicanism of the *Supper* has been the cause of considerable confusion; although it constitutes a major moment in Bruno's cosmological speculation, as well as offering important contributions to the larger sixteenth-century cosmological debate.[16]

There are four moments of *The Ash Wednesday Supper* which are directly concerned with discussing and developing the Copernican issue. They are the justly famous pages of praise of Copernicus in Dialogue 1; parts of the debate with the neo-Aristotelian scholar Nundinius in Dialogue 3; the final part of the debate with the other neo-Aristotelian scholar Torquatus in Dialogue 4; and the discussion of the movements of the earth developed by Bruno's mouthpiece, Theophilus, at the end of Dialogue 5. Some idea of the complexity of the context in which these separate sections are embedded can be gained from examining briefly the opening sections of the text which lead up to the pages in the first dialogue in praise of Copernicus.

15. For the various types of renaissance dialogue, see Cox 1992.

16. The major works on Bruno's cosmology to date are Michel 1962/1973 and the relevant sections of Vedrine 1967 and Ingegno 1978. For an account of the principal critical treatments of science and Copernicanism in Bruno from 1950 onward, see Maspero 1989.

Bruno's text opens with a poem in an anomalous sonnet form known as the "lame-sonnet" addressed to "The Malcontent," who turns out to be a jealous and sour-minded scholar who insists on clinging desperately to the traditional truths which the text will ruthlessly overturn. This is followed by a lengthy dedicatory letter to the French ambassador in London, Michel de Castelnau, Signor de Mauvissière, which reminds the reader that Bruno at the time of writing was serving in Castelnau's embassy as a gentleman attendant. The letter is divided into three parts. In a first section of remarkable literary refinement and virtuosity the polemical themes and moods of the ensuing dialogue are described through a series of exuberant negatives which only gradually give way to a positive definition of what the text will contain. The philosophical character of the work, when it is briefly referred to, is defined as Pythagorean: the only school that offers solid fare for the reader to digest. The second part of the dedicatory letter offers brief anticipations of the five separate dialogues that follow. The third and final part of the letter, again in a virtuoso mood, explains how the work aims to be all-inclusive, offering a comical-tragical portrait of the social and cultural context in which the multiple dialogues took place as well as delineating their contents. Here Bruno develops a theme present throughout his work: that of the text as a picture, which aims at including as many aspects as possible of the human and universal reality he is portraying. One of the principal difficulties in attempting to paint a cosmological picture, Bruno explicitly states, is that the human observer finds himself trapped in the picture, unable to take a few steps back to observe the form of the whole, as ordinary painters do, without falling into the abyss.[17] This concept was an essential factor in Bruno's skeptical stand, and it separates him from many of the more optimistic supporters and practitioners of the newly emerging sciences.

After these complex preliminaries the reader is taken into the first dialogue, which introduces in ferociously satirical terms the neo-Aristotelian scholars from Oxford, Nundinius and Torquatus. It was with them that the Nolan, during the supper organized by Fulke Greville, probably with the participation of Sir Philip Sidney, was called on to debate the Copernican cosmology. Even then the precise reference to Copernicus is deferred until the four participants have indulged in some apparently joking language games in the secondary dialogue through which the Ash Wednesday supper is recalled and discussed. The language games start

17. On the important connection between natural philosophy and the art of the painter, see Ciliberto 1990:67–70.

from the fact that, during the supper, the Nolan faced two opponents: Nundinius and Torquatus. This leads to a half-serious, half-comic *excursus* on the number 2, accompanied by a series of dichotomies, from the animals going into Noah's ark to the two Aristotelian principles of being, matter and form. The underlying purpose of this apparent digression is, however, clearly serious and important. Bruno is introducing the Copernican cosmological proposal in the context of a dialectical concept of being: a Pythagorean "play" of contraries whose final resolution into a higher unity lies outside the view of the painter of the cosmological picture. The cosmological picture itself is all that the human mind has to play with; and it is through that picture that Bruno proposes to reach an intuition of the infinite principle of unity which gives sense and meaning to the whole. The defining characteristic of Bruno's material universe, however, is its Pythagorean principle of duality; so that it is quite coherent, although it may seem surprising to a modern reader, that Bruno should precede his first explicit mention of Copernicus in his entire oeuvre by a remarkably sensuous and intensely poetic celebration of the English "muses". The sexual act, which Elizabethan London seems to have supplied him with in satisfying terms, furnishing his "pen" with a large and copious vein of prose, becomes the symbol of a cosmic ecstasy, eternally consumed and eternally repeated. This is Bruno rewriting the Platonic *Simposium* in heterosexual terms. At the high point of ecstasy, when he penetrates into the secret place covered by thirty seals and hidden in the darkness of deep shadows, the philosopher invokes the voice of memory in an attempt to recall the unity, or the divine principle of truth, from which the material cosmos derives. It is thus within the deepest cavities of his newly heliocentric universe that Bruno situates the possibility of knowing the first principles both of life and of truth.

At this point in Bruno's text, there is an apparently abrupt passage, from erotico-metaphysical to cosmological-physical explanation. The first mention of Copernicus is in the context of an invitation from a royal knight (in fact Fulke Greville, as earlier and later references make clear) desirous to converse with him on his Copernicanism and other "paradoxes" of his philosophy. The abruptness can, and I think should, be interpreted as a deliberate highlighting of the barrier separating metaphysics from physics: an awareness on Bruno's part that the new science was developing a new discourse that found and accepted its limits in the material world of quantity and measure. Copernicus's contribution to this discourse is immediately appreciated. Like Ptolemy, Copernicus, as a mathematical and observational astronomer, is among those who have "added light to light."[18] Bruno, however, immediately defines his own contribution in terms that clearly distinguish him from the practicing astronomers through a series of metaphors and quotations that lend weight and clarity to the distinction. The scientific astronomers are like translators who convert one language into another. They are like rustics who run to tell a captain of the way a battle is proceeding. They are like the Theban Manto who saw but failed to understand. With respect to all these examples, Bruno defines his inquiry as one of interpretation, rather than the gathering and systematization of facts. He is the one who understands the message which the scientists have "translated" from the language of the natural world. He is the captain who can make sense of the information given to him by the rustic observers. He is the blind Tiresius who can divine the true message behind the signs communicated to him by those who see.

These metaphors are far from being simple rhetoric or boasting, as they have often been interpreted. They are equally far from being the kind of distancing of himself, as a Hermetic Magus, from an empirical or mathematical science, seen as petty and bad, which Frances Yates considered them. On the contrary, Bruno is associating a true science with mental illumination and defining a world without it as a world of ignorance: a kingdom of those who see but are mentally blind. What he is doing, at the very beginning of the scientific revolution, is to separate off two distinct disciplines: that of the practicing scientist (which, in our post-Kuhnian era, we might want to call "normal science") and that of the philosopher of science. As a philosopher of science himself, Bruno recognizes his dependence on the data collected and ordered by the practicing scientists. At the same time, however, he claims his independence to comment on and interpret their discoveries in a far larger context than the one in which the "normal" scientist must operate, a context that will require the command of vision and imagination, as well as a prophetic insight that attempts to interpret nature in the higher terms of universal patterns or laws. Bruno's page can fruitfully be placed beside Bacon's later concluding remarks, in *The New Atlantis*, about the highest authorities of his House of Salamon: those who, albeit in more communal and less individualistic terms than Bruno's philosopher, nevertheless assume the

18. Bruno's insistence on knowledge as illumination participates in the development, so important in the emergence of the new philosophy, from the Augustinian idea of illumination as understanding of necessary and divine truths to the Cartesian idea of illumination as incorporated in a general theory of perception. On this subject, see Jolley 1990.

responsibility of ordering and "making sense" of the information supplied by the "normal" scientists:

> Then after divers meetings and consults of our whole number, to consider of the former labours and collections, we have three that take care, out of them, to direct new experiments, of a higher light, more penetrating into nature than the former. These we call Lamps.
>
> We have three others that do execute the experiments so directed, and report them. These we call Inoculators. Lastly we have three that [after consultation with the whole body] raise the former discoveries by experiments into greater observations, axioms, and aphorisms. These we call Interpreters of Nature.[19]

Bruno is suggesting that the new scientific inquiries are so important and far-reaching that the scientists themselves cannot be required or expected to draw out all the conceptual implications inherent in their discoveries about the universe. This is the task of philosophers of science. It is a surprisingly advanced intuition that looks forward not only to rational science-centered philosophies, such as those of René Descartes or John Locke, but also to the post-Kantian philosophies of the romantic period with their subjectively orientated and idealistic foundations. Furthermore, Bruno's interest in the technical aspects of the art of memory and the logic of Ramon Lull also relate him to an inquiry into the logic of scientific discovery, a theme particularly relevant to our own Popperian and post-Popperian age.

Only after Bruno has made these distinctions and clarified these aspects of his inquiry, is he ready to talk about Copernicus. In his first pages of explicit comment he is not concerned with the technical aspects of the Copernican astronomy but only with offering an overall appraisal of its achievements.[20] He begins by conceding copious praise to the genius of Copernicus, recognized in its profundity and maturity, and considered greater than those of preceding astronomers such as Ptolemy, Hipparchus, or Eudoxus. Then he introduces a critical note in claiming that, in spite of his gifts and industry, Copernicus did not go far enough in developing his cosmological model, for he left intact too many facets of the traditional Aristotelian-Ptolemaic universe. The reason for this failure is that Copernicus was too much of a mathematician and too little of a nat-

19. Bacon 1858–68, 5:411. The passage in brackets appears only in the Latin translation by Rawley.

20. For the cosmological and philosophical significance of Bruno's celebration of Copernicus, see the chapter titled "Copernico come segno divino" in Ingegno 1978:26–63.

ural philosopher. The criticism obviously intends to indicate that Bruno will be offering his own particular contribution to the cosmological debate, anchoring it to what he considers, in the physical world, "things which are constant and true." It is a judgment that would be reversed by Galileo in his later Copernican *Dialogue on the Two Major World Systems*, which claims that mathematics "knots" natural philosophy into certain and constant truths, relegating to metaphysical speculation questions such as the infinity of the universe which Bruno considered subject to logical proof.[21] For Bruno, the traditional mathematics held a purely abstract truth status and was considered as a departure from the physical realities of being rather than a true description of being itself. Having indicated the point at which he will take leave of Copernicus, Bruno then goes on to write a generous page of praise of the Polish astronomer's daring and coherent reversal of the traditional cosmology. Bruno follows Copernicus in considering the new heliocentric universe a rediscovery of "the ancient, true philosophy" rather than a discovery of something new. Furthermore, he gives full recognition to Copernicus for having contradicted the opinion of the gregarious multitude and having, on the basis of conclusions reached by his own "happy genius," raised himself up by a process of reasoning indicated by divine intelligence.

The remaining part of the first dialogue in *The Ash Wednesday Supper* is concerned with a definition of the Nolan philosophy rather than with Copernicus himself. It explains the terms of Bruno's own extension to infinite dimensions of the now heliocentric universe. The proposal was an audacious and polemical one within a social context which, as Bruno's Oxford experience had already indicated, was still having difficulties in accepting the more cautious heliocentricity of Copernicus. Bruno's awareness of the historical and cultural situation of Elizabethan London, and of the strength of the reaction his philosophy would be likely to meet even in an aristocratic circle like Sidney's where it could at least be discussed, is expressed in the allegorical journey through the dark, muddy highways of the nighttime city, which constitutes the second dialogue of the *Supper*. It is only in the third dialogue, when the Nolan is finally brought face-to-face with the first of his neo-Aristotelian opponents, Nundinius, that the Copernican cosmology is discussed in detail.

21. The relationship between Bruno's cosmological dialogue and Galileo's has been passed over in silence by recent generations of scholars, although it animated much comment between Tocco 1889 and Gentile [1920] 1991. However, the paper I gave on the subject at the conference "Galileo: Heretic Voices," Athens, September 1994, has been published in Athens in modern Greek, and has now appeared in English: see Gatti 1997 and also Aquilecchia 1995.

Nundinius proposes five subjects for debate, four of which are directly relevant, and agressively antagonistic, to the Copernican cosmology. The fifth proposal, which in chronological order is actually the first, is only apparently irrelevant to what will follow. Nundinius asks the Nolan if he understands English, and the Nolan replies that he has still not learned the language. A short discussion then follows about the diffusion of foreign languages in Elizabethan London whose purpose seems to be to establish that the cultured elite can readily communicate in Italian, thus justifying the Nolan's failure to learn English. Behind this apparent sociological digression, however, lies an important implicitly posed question: are all the characters involved in the debate speaking the same language? Or is the distance between them incommensurable? This question appears particularly relevant today when Thomas Kuhn's concept of linguistic and conceptual incommensurability in times of scientific revolution has been the subject of such a heated debate.[22] A conversation does develop, however, between the guests at the Ash Wednesday supper, albeit with much ill humor and a large dose of mutual misunderstanding.

The four properly Copernican subjects, in the order proposed by Nundinius, are, first, that Copernicus was not putting forward a realist astronomy which made the earth actually move, but only a mathematical hypothesis that would save the phenomena more exactly than in the Ptolemaic system; second, that it is impossible for the earth to move because it is the fixed center of the universe and the foundation of all that moves around it; third, a question is asked concerning the quality of the matter in those bodies which the traditional cosmology claimed to be made up of a celestial quintessence; fourth, that if the earth were really to move in an eastward direction, all the clouds would appear to stream toward the west. The third and central dialogue of Bruno's text concerns the Nolan's answers to these proposals.

The answer to the first proposal takes up a large part of the dialogue. Bruno's insistence on its importance indicates his awareness of the way the cosmological debate had been developing since the publication in 1542 of Copernicus's *De revolutionibus*. Historians of science, particularly Robert Westman, have demonstrated how the Wittenberg school of Copernicans accepted and developed the mathematical aspects of the new cosmology while refusing to contemplate its realist implications.[23] Given that this reply to the Copernican proposal was especially con-

22. For a discussion of this much-debated question see Gabbey 1990; for bibliographical indications of the *loci classici* of the discussion, see ibid., 521–22, n. 9.

23. See Westman 1975a.

nected with the Protestant culture of Wittenberg, it is no surprise to see it being repeated by the Oxford-based Nundinius as his principal weapon against the radically realist interpretation of the Nolan. Bruno's reply to this evident compromise with theological doctrine is both lengthy and complex. Like the rest of the supper, it is not related by the Nolan himself but by the figure of Theophilus, who claims to have been present at the supper as one of the Nolan's most faithful disciples.

Bruno's answer to Nundinius rightly takes as his starting point what he calls the "epistola superliminare" to Copernicus's book: the anonymous letter prefixed to *De revolutionibus* by the Protestant theologian Andreas Osiander while Copernicus was on his deathbed. Osiander's letter first proposed the Copernican astronomy as a "treasure of suppositions aimed only at an admirable and artificial improvement in calculation." Bruno translates a large part of the letter into Italian and reproduces it in his text. He does not name the author, whose identity he may not have known, but he claims (and was the first to do so in print) that it could not have been written by Copernicus, for it is in direct contradiction with Copernicus's dedicatory letter to Pope Paul III which calls on the ancient Pythagoreans and Timaeus to support a realist heliocentric universe.[24] Copernicus suggests to the pope that such a universe may be outside the comprehension of the masses and should remain the carefully guarded secret of a cultured elite; but this proposition is very different from that made by Osiander. We have already seen Bruno's reply to such a proposition in our discussion of his Pythagoreanism, and his public and published debate on the Copernican cosmology in this text is in line with the conclusion of Dialogue 1, which advocates the careful diffusion of the new forms of knowledge among all who are interested in learning about them.

Having concluded his general attack on Osiander as a deceitful and dishonest guard of Copernicus's new cosmological edifice, Bruno goes on to consider a specific optical argument which in Osiander's opinion proves the physical impossibility of the heliocentric hypothesis, that is, that Venus acquires in the new astronomy an angle of 40 degrees or more on either side of the sun. Osiander argues that if this angle derives from the movement of Venus on an epicycle while she orbits around the sun, then, in terms of Euclidean optics, the size of Venus will appear sixteen times as large when she is nearest the earth than when she is furthest away: an obvious absurdity in contrast with experience. Bruno replies by pointing

24. Although Bruno was the first publicly to denounce Osiander's letter as spurious, it was Kepler who first revealed the name of its author, in his *Astronomia nova* of 1609.

out that a "true" optics will take into consideration factors not considered by Osiander, such as the degree of luminosity of the body concerned and the fact that at great distances the apparent sizes of luminous bodies appear constant. By a "true" optics, Bruno probably meant the optics of Alhazen (Ibn Al-Haytham), the eleventh-century Arabic astronomer whose book, known in Latin as the *Perspectiva*, was read with enthusiasm by the ninth earl of Northumberland, who brought together in his library at Petworth House one of the major English collections of Bruno's works.[25] In book III, chapter 7, of his text, Alhazen considers "The Ways in which Sight Errs in Inference" and writes that "by looking at a fixed star and a planet at the same time sight will not perceive the difference between their distances, but rather perceive them both in the same plane despite the great difference between their distances".[26] Bruno also wishes to prove that opaque bodies lying between the eye and luminous bodies can easily disappear from the field of vision: a point which he illustrates by holding a matchstick between his eyes and a lighted candle. This deceptively simple piece of reasoning constitutes an essential premise for the new cosmology of Bruno, whose universe is filled with infinite worlds, most of which are invisible to the naked eye. Even in the dimensions of the visible universe, Bruno is convinced of the existence of unseen worlds, perhaps even relatively close to the earth.[27] It was this rather sophisticated intuition that would make Kepler chide Galileo, on the publication of his *Starry Messenger* announcing his telescopic observations of the so far unseen moons of Jupiter, for failing to acknowledge any debt to those precursors who had helped him toward his discovery, including his own countryman from Nola.[28]

25. Northumberland mentions his reading of Alhazen as the factor that changed his life from that of a frivolous courtier to that of a dedicated natural philosopher, in his autobiographical essay first published by Frances Yates with the title "On Love" as an appendix to her study of *Love's Labours Lost*. See Yates 1936:206–11. For the Northumberland collection of Bruno texts, see Gatti 1989:35–48. On the importance of Alhazen's work for renaissance thinkers, see Field 1997:7.

26. See the relevant pages in Alhazen 1989.

27. There may also be a silent reference here to the work of Jean Penel (or Pena), who had attempted in his *Euclidis optica et catoptica*, published in Paris in 1557, to prove the possible movement of the earth through optical reasoning. For Bruno's probable use of Pena, see Vedrine 1967:218 and Sturlese 1985:317–19. For the importance of Pena's work in demolishing the solid celestial orbs, see Goldstein and Barker 1995.

28. Kepler made his complaint verbally to Martin Hasdale, librarian to the imperial court at Prague, who conveyed it to Galileo in a letter of April 15, 1610. The letter, which is in Italian, was published in Galilei 1890–1909, 10:314–15.

Nundinius's second Copernican proposal receives very brief treatment by Bruno. As Theophilus immediately points out, Nundinius is presupposing his own central proposition: that the earth remains fixed at the center. The Nolan thinks otherwise; and the logical clumsiness of Nundinius's proposal here gives Theophilus a chance to develop his own idea of an infinite universe in which no body can claim to be at the center except in terms relative to a particular point of observation. At this point, Copernican heliocentricity has clearly been surpassed in a far more complex and innovative cosmological picture. Nundinius, in dismay, seems unable to propose anything further except a question about the nature, in the Nolan cosmology, of those bodies traditionally considered quintessential. Theophilus is quick to point out that this question too is logically unacceptable because it jumps from the main issue of the debate, which is Copernican heliocentricity, to a related but peripheral issue.[29]

Nundinius's question about the matter of which the infinite bodies in an infinite universe would be composed was of great importance in Bruno's philosophy, which denied the basic Aristotelian assumptions about the static and quintessential nature of celestial matter. Here, however, he gives Nundinius's question short shrift, possibly wishing to give his reader only a taste of a difficult subject he would take up in far greater detail later on. Another possibility is that Bruno's hasty answer to this question is motivated by memories of his previous Oxford disgrace. If his lectures there really had been interrupted after a reference to the magnetic theory put forward by Ficino in chapter 15 of the *De vita coelitus comparanda*, which suggested that magentism was a heavenly as well as an earthly phenomenon, Bruno may have been hesitant about insisting again too explicitly on such a delicate subject. "The Nolan" actually proposes his ideas on this question, at this point of the debate, as possibilities rather than certainties, although the ideas put forward are those which Bruno maintained until the end of his philosophical inquiry. Briefly, he sees the entire infinite universe as being inhabited by two kinds of bodies: earths and suns, defined as cold and hot bodies. That is, all bodies in the universe are composed of varying forms of the same ba-

29. Bruno explicitly states, in the *Supper*, that his intention is not to abandon Aristotelian logic altogether, but rather to use his opponents' weapon against them to prove the errors in their reasoning. This stand explains what would otherwise be the rather perplexing statement found in the *Marginalia* of Gabriel Harvey (1913:156): "Jordanus Neopolitanus, (Oxonij disputans cum Doctore Underhill) tam in Theologia, quam in philosophia, omnia revocabat ad Locos Topicos, et Axiomata Aristotelis; atque inde de quavis materia promptissime arguebat." On this important note see Aquilecchia 1995:26–28.

sic, infinite substance, and no essential difference exists between sublunar and heavenly bodies. Everywhere earths circle around suns in order to receive their life-giving heat and light.

Undoubtedly Bruno has in mind here the thermodynamic philosophy proposed in these years by his Neapolitan fellow natural philosopher Bernardino Telesio, who saw heat and cold as the basic dichotomy that determined universal vicissitude. Bruno knew and admired Telesio's work, even if the Telesian universe, from a cosmological point of view, remained within the Aristotelian-Ptolemaic scheme.[30] Nevertheless, Telesio's universal warring substance was an important element in defining Bruno's newly infinite cosmological perspective according to which all the earths in the universe are substantially the same as our earth, just as all the suns are substantially the same as our sun. Having proposed this distinction, however, Bruno goes on to deny any essential difference between hot and cold bodies by pointing out how the earths, while they receive heat from their suns, nevertheless then nurture within themselves, and are even capable of reflecting outward, the heat they have absorbed. Bruno sees the need for heat on the part of earths, and a corresponding need to communicate heat on the part of suns, as the principal cause of the celestial motions: an important point, given that his radical rejection of the Aristotelian-Ptolemaic closed spheres left his universe without a prime mover to put in action the now superfluous celestial orbs.

After his repeated failure to find persuasive and logically valid arguments with which to forward his defense of Ptolemy, Nundinius finally gets back to the mainstream of the argument with a proposition that had traditionally been considered a bastion of the geocentric hypothesis. That the clouds would always move to the west in the case of an earth that every twenty-four hours revolves around itself eastward was a standard defense of a stationary earth which had been used by Ptolemy in the *Almagest*.[31] Following Copernicus, Bruno replies to this argument with quotations from both Plato and Aristotle's *De meteorologia* which suggest that the earth is to be seen as including a first layer of its atmosphere, in which the clouds move together with their accompanying earth as if they were part of its being. Bruno uses the metaphor of the earth as an animal, whose deep caverns and seabeds he sees as the dark humors of the liver and whose clouds and atmosphere he likens to lungs. The outer cir-

30. On Bruno and Telesio, see Aquilecchia 1993a, sections XX and XXI, and Gatti 1994b.

31. See *The Almagest*, I, 7, H25–26, in Ptolemy 1984.

cumference of the globe of the earth, according to this concept, should not be drawn around the surface of the solid globe but around the limits of its accompanying air and atmosphere.[32]

Having reported the Nolan's reply to this first traditional objection to heliocentricity, Theophilus is then asked by the Englishman Smitho to explain a related and more demanding phenomenon: why a stone thrown onto a ship from the bank of a river will miss its target, because the ship will have already proceeded onward, although a stone thrown from the mast of the ship will always end up at the foot of the mast and not behind it. This also was a traditional argument in the context of a discussion of a moving earth, and Bruno's excellent and justly admired reply in terms of a theory of impetus closes the third dialogue of the *Supper* on a confident and strongly pro-Copernican note. Bruno's argument is actually far stronger than that put forward by Copernicus himself to solve this point; for Copernicus only extended to this argument the idea of a unity and sympathy between the ship and the stone, just as there is a unity and sympathy between the earth and its atmosphere. The concept, however, remained vague, as it is not clear in Copernicus's text exactly what kind of sympathy links the stone to the ship. Bruno (although he fails to mention them) is basing his more adequate reply on the impetus theories of the medieval Parisian philosophers such as Nicole Oresme and Jean Buridan. Bruno's use of the impetus argument in this page of the *Supper* suggests that he was well versed in the fourteenth-century Parisian discussion of celestial mechanics. The terms of this debate were made clear to the twentieth-century reader by Pierre Duhem's publications of the essential manuscripts involving texts of Franciscus de Marchia (*On the Sentences of Peter Lombard*, ca. 1320), Albert of Saxony (*Subtilissimae quaestiones in libros de Caelo et Mundo*, 1368), and, in the latter part of the century, the discussions by Jean Buridan and Nicole Oresme of Aristotle's *Physics* and his books *On the heavens*.[33] All these authors, however, were discussing the possibility of a rotation of the earth around its own axis, rather than of the sphere of the fixed stars around the earth, and none of them questioned seriously the central position of the earth in the universe.

One of the central points in this early Parisian discussion was precisely the Aristotelian objection raised by Torquatus, in Bruno's dialogue, to the

32. See Aristotle, *Metereologica*, I, 3, and Plato, *Phaedo*, 109, b–e.

33. For the impetus theories of Buridan and Oresme, see Clagett 1961. At pp. 665–66 Clagett mentions Bruno's use of the medieval theories of impetus.

movement of the earth: that is, that an object thrown up vertically into the air would fail to fall to the same place if the earth were in movement. Both Buridan and Oresme use the ship image taken up by Bruno to illustrate the relativity of moving objects in space, and in doing so they arrive at the idea of closed mechanical systems of movement related in space. Buridan, in his comment on Aristotle's *Physics*, also developed a theory of impetus to account for moving projectiles: "Thus we can and ought to say that in the stone or other projectile there is impressed something which is the motive force of that projectile."[34] Buridan himself fails to apply his impetus theory to an explanation of the vertical drop of the stone from the mast of the ship; but Oresme does precisely that, in his comment on Aristotelian cosmology, in a page that reads remarkably like Bruno's. Marshall Clagett and others have demonstrated how the echoes of this fourteenth-century Parisian discussion were still distinctly present in the sixteenth-century, post-Copernican cosmological debates; and this page of Bruno's has been placed in a line of development that passed from the *Supper* through Kepler to Galileo.[35] It would be a mistake, however, to overemphasize Bruno's dependence on these medieval precedents. He is applying the impetus theory to a heliocentric, post-Copernican universe undreamed of by any of the scholars mentioned above. Bruno's concept of relative mechanical systems thus becomes one of his most advanced scientific intuitions and justifies his claim that he is not simply concerned in this dialogue with reproposing the arguments of Copernicus. Bruno's mechanical and nonmathematical treatment of the heliocentric hypothesis testifies to his awareness, which has been the subject of some eloquent pages by Alexander Koyré, of the need, in the wake of the Copernican revolution, to reformulate the whole study of physics.[36] Galileo would later carry out this program on the level of experimental physics. Bruno, for his part, remains within a philosophical and theoretical context of argument, although not unheeding of the need to proceed toward observation and measurement. His dialogue opens up the question of relative frameworks of motion in a universe that has lost forever the unique point of reference supplied by a central earth in the Aristotelian-Ptolemaic cosmology.

The fourth dialogue of the *Supper* brings the Nolan face-to-face with the second of his Oxford neo-Aristotelian opponents, Torquatus. This debate is far more disordered and fragmented than that in the third dialogue,

34. *Questions on the Eight Books of the Physics of Aristotle*, para 4. See ibid., 534.
35. See also Wallace 1981.
36. See Koyré 1966.

which was structured around the five proposals of Nundinius. Bruno uses this fragmentation to his advantage, addressing a series of larger but closely related issues such as the fraught question, which will become of such primary importance to Galileo, of the scarce biblical authority for a heliocentric universe. Bruno here anticipates an argument that Galileo would use later on: that the Bible should be considered as moral teaching for the masses and not as a scientific-cosmological text at all.[37] It is only at the end of the dialogue that the Copernican issue is again taken up in pages of essential importance for a correct understanding of Bruno's reading of Copernicus; for it is here that we find what has traditionally become known as "Bruno's Copernican Mistake."

This surprisingly ignored episode of *The Ash Wednesday Supper* has received attention above all from Frances Yates and the followers of her Hermetic interpretation of Bruno, for it appears to supply them with proof that Bruno was no technical astronomer, and really no Copernican at all.[38] Such a reading seems to me mistaken. Although it is clear that Bruno is refusing to play the role of faithful disciple and is not offering a conventional reading of Copernicus's astronomy, there seems no reason to deduce from his effort to extend the Copernican universe to infinite dimensions that his journey through the *De revolutionibus* had been hasty and inattentive or that his infinite universe does not have a properly astronomical and physical foundation. It was certainly a difficult journey, based on what Bruno himself recognizes in the later *De immenso* as a fragile technical training. Sometimes he slipped up on technical detail, but so did most of his contemporaries. Copernicus's book, as Bruno himself points out in this same dialogue, was not easy to read in part because of his insistence on continuing to use a Ptolemaic vocabulary instead of reforming his terminology together with his cosmology. Nevertheless, Bruno recognizes in *The Ash Wednesday Supper* that Copernicus had changed his life. There was a time "when I was a stripling," he writes in

37. Bruno's use of and attitude toward biblical knowledge is still largely unexplored. On the one hand, he is clearly disapproving of too close and dogmatic a reference to biblical texts, particularly in the context of natural philosophy. On the other, his experience in the Protestant part of Europe seems to have stimulated a more independent and direct reading of the biblical books than the Catholic world allowed. For Galileo and the Bible, see Blackwell 1991.

38. Yates 1964:237–38: "This ancient truth now reviving of which the Nolan is the prophet is not heliocentricity in an astronomical sense, or as a mathematical hypothesis. This is the clue; the vision which the Nolan is developing is a new insight into the divinity of the universe, an expanded gnosis." This judgment is echoed in Bruno 1977:26, where the translators claim that *The Ash Wednesday Supper* is concerned with the Copernican theory only in a subsidiary and metaphoric sense.

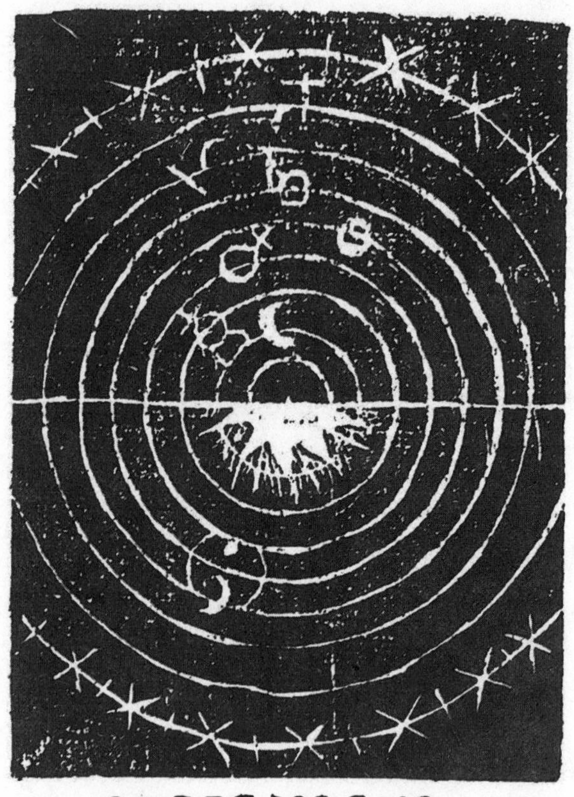

2. Diagram in Giordano Bruno, *La cena de le ceneri* (1584), dialogue 4. Reproduced by permission of the Director of Library Services, University College London.

the fourth dialogue of that work, when he thought the Copernican hypothesis was a joke, put forward for the pleasure of disputing an obvious sophism. Then his ignorance and blindness were healed, and he entered into his maturity as a philosopher in full recognition of the truth of the Copernican proposal.[39] These are surely not the words of a thinker for whom reading the *De revolutionibus* had been a simple or marginal task.

39. Bruno 1958:136; Bruno 1977:187–88.

Let us get this usually misread page of the fourth dialogue of *The Ash Wednesday Supper* into perspective.[40] Irritated by the Nolan's refusal to accept the continuing superiority of the Aristotelian-Ptolemaic astronomy, Torquatus calls for pen and paper and draws an elementary figure representing on the upper half of the sheet the Ptolemaic universe and on the lower the Copernican: a placement of the two systems which is already a comment on Torquatus's conviction of the "superiority" of Ptolemy with respect to Copernicus. Torquatus then starts to illustrate the Copernican system, only to be interrupted by the Nolan at the point where he claims that the moon is carried around the sun on an epicycle whose center is the earth. The Nolan claims that the point at the center of the epicycle, which is considered by Torquatus as the earth, is nothing more than the mark made by the foot of the compass when the illustration was drawn. According to the Nolan, Copernicus placed both the earth and the moon on the circumference of the epicycle. He gives as his reason that if the earth were at its center, and therefore at an unvaried distance from the sun throughout the year, the diameter of the sun would always appear the same. Copernicus's book is then called for and, as Bruno's mouthpiece, Theophilus (who narrates this episode in the text) admits, the illustration of the new universe does not place the earth on the circumference of the epicycle, as the Nolan claimed, which is why Torquatus interpreted the point at the center as the earth.

Nevertheless, Bruno once again, in the figure of the Nolan, accuses Torquatus of an erroneous reading of Copernicus, claiming that both the earth and the moon are, in his written text, placed on the circumference of the epicycle. For Copernicus, he claims, wrote that the earth and the moon are contained within the same epicycle. This is a reference to Copernicus's own words on the page of book I where the diagram of the universe was printed, words which Bruno clearly interprets as meaning that Copernicus put both the earth and the moon on the circumference of the epicycle. Bruno's own diagram of the Copernican universe, in first editions of the *Supper*, clearly reproduces this model (see Fig. 1).

Bruno's continuing insistence that Copernicus's astronomy put the earth on the circumference of an epicycle is what is generally considered his Copernican mistake. There is no doubt that it contradicts the illustration of the new universe given in book I of Copernicus (see Fig. 2). Copernicus's English translator and editor, Edward Rosen, has pointed out, however, that the illustration of the new universe in the first edition

40. The pages considered here are in Bruno 1958:139–41; Bruno 1977:190–93.

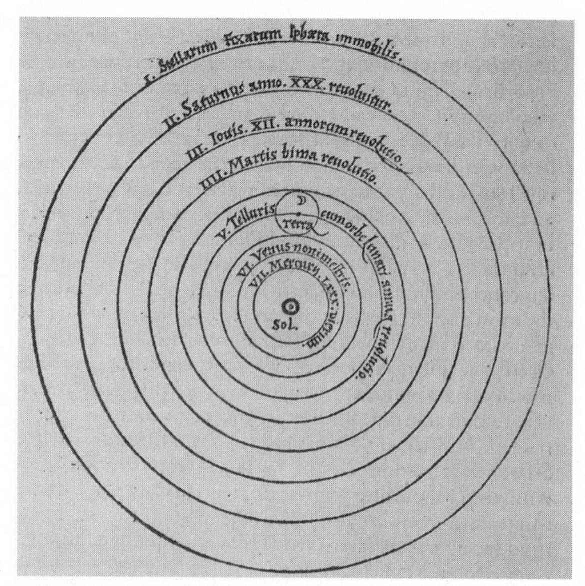

3. Diagram in Copernicus, *De revolutionibus* (1543), book 1. Reproduced by permission of the Director of Library Services, University College London. Copyright © Library Services, Education and Information Support Division, University College London.

of *De revolutionibus*, published in 1543, and in the second edition of 1566—one of which is being consulted in Bruno's *Supper* in 1584—was not based on Copernicus's autograph drawing but on an imperfect copy which, in Rosen's opinion, may justifiably be considered the cause of considerable confusion.[41] Copernicus's manuscript drawing, which Rosen publishes in his edition, was admirably clear and concise. Significantly, however, it failed to give a graphic representation of the earth's or the moon's orbits but simply carried the words written inside the orb of the earth: "*Tellurus cum Luna annua revolutio*," translated by Rosen as "The annual revolution of the earth, together with the moon."[42]

Before going any further into the question of the diagrams in Copernicus's book, I wish to clear up a philological problem concerning the reading of the relevant passage in book I of Copernicus's text. Bruno is claiming that the diagram published in the first editions of Copernicus's text was inaccurate and that Copernicus never intended to put the earth at the center of its epicycle carrying the moon, with both of them orbiting around a central sun. Bruno backs up this claim by referring to the pas-

41. See Copernicus 1543/1978:359.
42. Ibid., 21.

sage in book I of *De revolutionibus* where Copernicus writes: "*Quartum in ordine annua revolutio locum obtinet, in quo terram cum orbe lunari tanquam epicyclio contineri diximus.*"[43] Bruno translates this into Italian as "*Lessero e ritrovarno che dicea la terra e la luna essere contenuta come da medesmo epiciclo.*" The English equivalent to Bruno's translation would be "They read and found it said that the earth and the moon are contained as in the same epicycle." Hélène Vedrine claims that Bruno's translation of Copernicus's Latin is badly mistaken. She thinks that "tamquam epiciclo" can only refer to the moon, and not to the earth and the moon together as in Bruno's translation.[44] Frances Yates was of the same opinion. Indeed, in *The French Academies in the Sixteenth Century*, Yates had pointed out that Pontus de Tyard, in his French translation of Copernicus's text published in 1552, titled *Discours des parties de la nature du monde*, had also rendered this passage in the same sense as Bruno. Tyard's translation runs: "*come dans un épicycle, la Terre et toute la région Elémentaire avec le globe de la Lune est contenue.*"[45] It would seem an extraordinary error in translation for two such distinguished humanistic scholars to make, especially as Copernicus had written shortly before that everything which the moon "embraces" such as the center of the earth (*totum, quod Luna praecingit, ac centrum Terrae*) passes around the great orb with an annual revolution around the sun, and around the sun lies the center of the world. The problem may be that Copernicus deliberately keeps his text somewhat ambiguous, just as he deliberately refrains from being too specific in his own diagram. Both the text and Copernicus's diagram become much more understandable if reference is made not only to the passages in book I of *De revolutionibus* which are in question but also to the pages of book III where Copernicus returns at length to the problem of the precise relation of the movements of the earth with respect to those of the sun and the moon. For in book III it appears that Copernicus himself was not sure about some aspects of this issue.

Bruno correctly defines the problem in his objection to the confusing diagram of the universe in book I of the first editions of *De revolutionibus*. In that diagram, in addition to the moon's perihelion and aphelion,

43. Copernicus 1543:f. 9r and v. Some commentators claim that Bruno and his fellow guests at the supper were looking at the edition of 1566, which exactly reproduced that of 1543 but contained also the *Narratio prima* by Copernicus's disciple Rheticus. See in particular Granada 1991. The copy of Copernicus's book with the name "Brunus" written on the flyleaf belongs to this 1566 edition.

44. See Vedrine 1967:231.

45. See Yates 1947:102–3, n. 5, and Tyard 1578:99.

which also mark the aphelion of the orb of Venus and the perihelion of the orb of Mars, the diagram contains a separate circle for the earth's center, so that the earth-moon system lies through three circles instead of the usual two (see Fig. 3). That is, the earth (if one accepts that it corresponds to the point at the center of the epicycle) appears to lie permanently at the center of its orb, and its orb moves around a central sun. Bruno is right to claim that if the earth moved around the sun in a perfect circle, as it appears to do in the diagram and according to Torquatus's interpretation, we would always see the sun with the same diameter. That the sun's apparent diameter varies considerably is defined by Copernicus himself, in more complicated terms, as the "nonuniformity of the sun's apparent motion." In book III, chapter 15, of *De revolutionibus* he claims that there are two possible ways of defining this nonuniformity graphically and mathematically. One is by means of an eccentric, so that the earth revolves in an annual orbit around a sun that lies at some distance from the geometrical center. The other is by means of an epicycle on a concentric, with the earth moving around the circumference of an epicycle which itself revolves around a central sun. Copernicus, in book III of *De revolutionibus*, merges these two answers to the problem in a single diagram to accentuate the fact that they represent two alternative ways of visualizing the same phenomenon: the annual revolution of the earth around the sun (Fig. 4).

If we now come back to the general diagram of the new universe in book I of Copernicus's text, printed in the first editions (Fig. 3), we can see that Torquatus, in interpreting the earth as the point at the center of the epicycle, is thinking of the first of these solutions put forward by Copernicus in book III, chapter 15, that is, the solution of a simple orbit on an eccentric. The diagram represents this solution imperfectly, however, because it places the sun at the geometrical center of the universe instead of on an eccentric. Bruno himself is thinking of the second solution proposed by Copernicus in book III, chapter 15: that of the epicycle on a concentric. But Bruno's interpretation of this model (including his diagram in book IV of the *Ash Wednesday Supper*; see Fig. 2) is also incorrect as in it he identifies the epicycle of the earth (recognized by Copernicus in book III, chapter 15, as one way of explaining the earth's motion with respect to the now central sun) with the epicycle of the moon.

It is clear that the moon should have been put on a further epicycle around the earth. Copernicus himself discusses the moon's motion around the earth in book IV, chapter 3, of *De revolutionibus* and, in the context of the epicyclic solution to the earth's motion, finishes up by putting the moon on a second epicycle, or epipicycle, centered on the

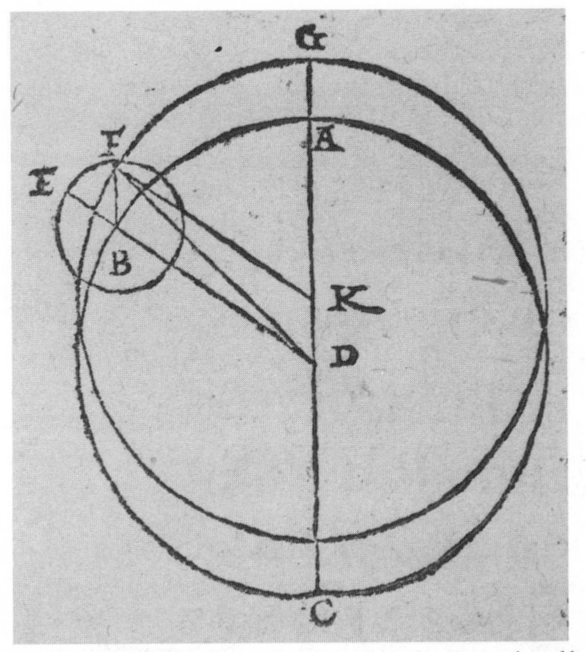

4. Diagram in Copernicus, *De revolutionibus* (1543), book 3. Reproduced by permission of the Director of Library Services, University College London. Copyright © Library Services, Education and Information Support Division, University College London.

earth: a solution which, as his near contemporaries such as Tycho Brahe would be quick to point out, rendered immediately obsolete the idea of solid orbs which Copernicus himself never denied. For the moon's epicycle, at some points, would clearly penetrate the space of the first epicycle and shatter its crystalline perfection.[46] This contradiction, however, would not have worried Bruno, whose theory of matter denied the celestial quintessence of the traditional cosmology in favor of a unique substance stretching out into infinite space. If Bruno mistook the movement of the moon around the earth, putting them both on the same epicycle rather than adding another epicycle, or epipicycle, for the moon centered on the earth, he had other reasons. The principal reason is probably to be found in his thermodynamic explanation of the movement

46. Brahe's repudiation of the solid celestial spheres was published in his *De mundi aetherei recentioribus phaenominis* in 1588. Bruno so admired Brahe's work that he dedicated one of his books to him, but the admiration was not returned; for theological reasons, Brahe could not accept the infinity of Bruno's universe. For Bruno and Brahe, see Sturlese 1985, and for the importance of the astronomical works published in 1588, including the *Camoeracensis acrotismus* of Bruno, see Granada 1996.

of the worlds, which furnished a reason why the planets should revolve around the sun but failed to explain why a cold moon should revolve around an equally cold earth. In terms of Bruno's explanation of planetary motion, his solution of putting the moon on the same epicycle as the sun is perfectly reasonable.

The first aspect of the question to be noticed is that this lunar mistake is quite different from the one traditionally imputed to Bruno. Usually he is considered to have misunderstood Copernicus by putting the earth on the circumference of an epicycle moving in a circular orbit around a central sun, a solution not mentioned by Copernicus in book I of his text but admitted at least as a mathematical possibility in book III. Perhaps the problem is that Copernicus never distinguished clearly between mathematical and realist aspects of his astronomy, a failure for which Bruno repeatedly chides him. Bruno's mistake derives from an incapacity to understand how the choice of putting the earth on an epicycle affects the orbit of the moon. His erroneous solution to this problem clearly indicates that he had no idea that the moon was the satellite of the earth or, more serious still, that it was necessary to account accurately for the phases of the moon during its monthly cycle. In his Copernican dialogues Bruno offers no astronomical tables or mathematical explanations, although at the end of the first dialogue of the *Ash Wednesday Supper* he does express his admiration for the increasing observational skill of professional astronomers throughout history. Furthermore, in his later comments on the Copernican theory in the third book of *De immenso*, Bruno returns at length to the question of the moon's movements around the earth. At that point he was no longer satisfied with his own solution and also realized that he had mistaken the solution to the monthly phases of the moon. Unfortunately, he could find no better answer to his dilemma than to accept the proposal of Torquatus which he had refused in the *Supper*. He now offers his reader a long explanation of this solution, which repeats the mistake of making the earth circle around a central sun, just as Torquato had argued, rather than around an eccentric. Bruno even confuses the issue further by excogitating a system according to which Venus moves in an epicycle around Mercury, as a kind of pendant to the earth-moon system.

There may be some sense, nevertheless, in Bruno's lunacy, to make a pun. For it is clear that his problems partly arise not so much because he misunderstands Copernicus's arguments as because he dislikes the whole system of epicycles and eccentrics on which Copernicus's astronomy, like Ptolemy's, was based. In the *De immenso* he explicitly classifies both eccentrics and epicycles, with the celestial orbs themselves, as fantasies and

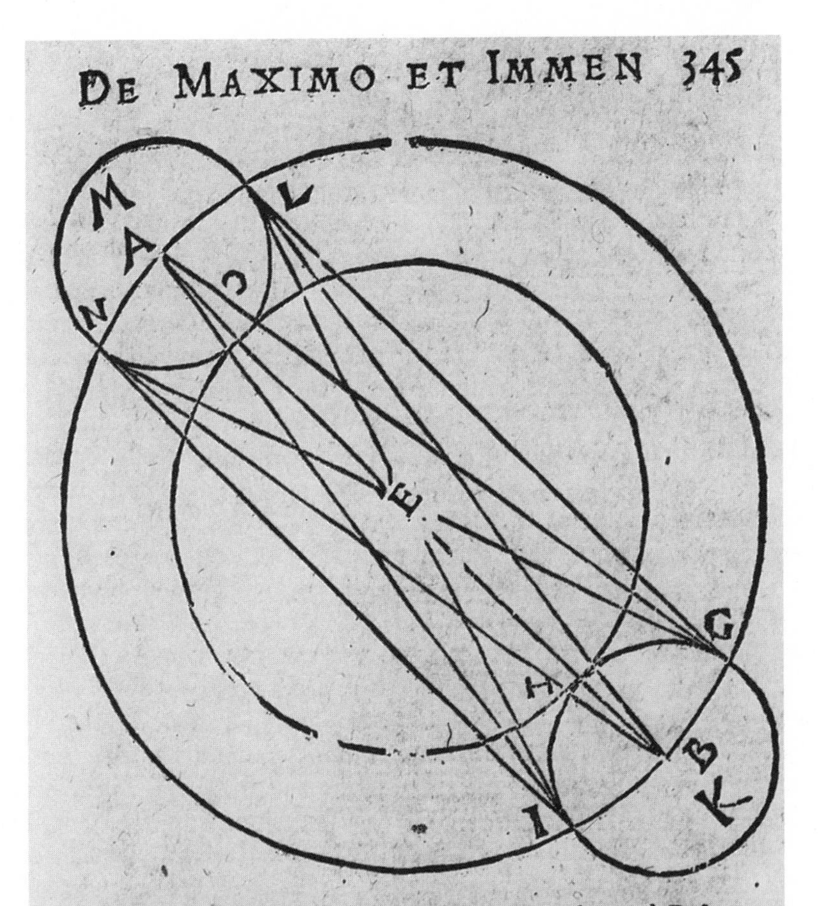

quas deſcribũt rectæ hinc A B, A I, A G, inde verò B A, B L, B N; Tellus & Mercurius viciſſim à corpore ſolis minore videãt elongatione feriare? hinc verò K Venere inde vero M Luna magis à ſole recedentibus. quod in hoc ſchemate, ut compẽdioſiore, ita & cõfuſiore ratione deſignamus quãdoquidem neq; facile eſt nobis circulorum multiplicatione vel (ut reuera naturæ conſentaneum) ſpiralium, fælicem compoſitionẽ ut poſsimus adducere, præſertim cũ vix in reb. ad ut phyſicum hoc inſtitutũ aperiẽdũ multũ nobis adſit ocij: ideo ſolertiorib. Aſtronomis rei tãtæ ita mõſtraſſe locũ ſufficiat, ut ipſi per ſe ipſos melius quam ego indicare poſsim reliqua perſequantur, ubi circa A phænomenas telluris & lunæ, circa B Veneris & Mercurij, circa E ſolis inquirẽ-

Y 5 das ſi-

5. Diagram in Giordano Bruno, *De innumerabilibus, immenso, et infigurabili* (1591), book 3. Reproduced by permission of the Biblioteca Nazionale Centrale, Rome.

chimeras which astronomy would do well to eliminate. Bruno may, like Kepler, have mediated this dislike through Proclus, whose *Hypotyposis* expresses an intense distrust of a complicated mathematical astronomy and a particular dislike of epicycles.[47] The fact that Bruno lived before Kepler's discovery of elliptical orbits offered the possibility of calculating the movements of the planets without the need for eccentrics or epicycles does not diminish the importance of his realization that, like the revolving celestial orbs, they were purely conceptual entities that rendered a translation into physical terms of the new astronomy an arduous and risky task. Thus it may seem strange that Bruno, when faced with Copernicus's alternatives of the eccentric and the epicycle, both of which fairly correctly represent, in mathematical terms, the relative movements of the earth around the sun, chose, in the first instance, the more complicated epicycle rather than the eccentric. Copernicus, in book III of his text, made no definite choice between the two models but only claimed that it was difficult to know which might be the true one. Bruno, in *The Ash Wednesday Supper*, also gives no reason for his choice of Copernicus's alternative solution which puts the earth on an epicycle with respect to a central sun. In Book I, chapter 5, of the later *De immenso*, he does justify his choice. The sun, he says there, is central, and does not appear to vary in its diameter because of an eccentric orbit of the earth, which he evidently thinks of as a purely mathematical expedient. The reason is rather that the dynamic impetus of the earth's soul carries it around on its epicycle together with the moon, just as all the other worlds circle freely around their central heat-giving suns.[48] Bruno always maintained a biological-animistic approach to cosmological questions, favoring any technical solution that guaranteed his idea of a universe involved in a life-giving process of constant vicissitude and metamorphosis. In *The Ash Wednesday Supper* he accuses Copernicus, with some justice, of being too much of an abstract mathematician and of failing to provide physical arguments justifying choices between differing mathematical models of the universe. Bruno himself can be accused of putting physical arguments too emphatically at the center of his astronomical speculation, appearing at times to ignore the necessity of making his theory coincide with observation and calculation.

Nevertheless, it is fair to claim that Bruno is raising a point which Copernicus himself had had difficulty in dealing with and which Bruno's neo-Aristotelian opponents (at least in the terms of Bruno's dialogue)

47. See Field 1988.
48. Bruno 1879–91, I, i:218–19.

seem incapable of discussing with him. Maliciously, and perhaps a little dangerously, Bruno claims that they have not read Copernicus carefully enough. It is nevertheless fair to point out that difficulties in interpreting technical niceties of the new astronomy lasted well into the seventeenth century. On June 20, 1612, the young Prince Federico Cesi, who had just welcomed Galileo as the most distinguished member of his Accademia dei Lincei, wrote Galileo a letter in Florence asking if he thought Copernicus proposed two or more epicycles to explain the phenomena and if he still thought in terms of solid celestial orbs. Galileo replied on June 30, 1612, that Copernicus used one or other hypothesis indifferently, and no longer believed in solid orbs.[49] In fact, not everyone today would agree that Copernicus had done away with the solid orbs as easily as Galileo surmises. His correspondence with Cesi (together with the intricate discussion that later developed around Galileo's own *Dialogue*) shows that reading Copernicus's text correctly was still a problem well into the seventeenth century.

The fifth and last dialogue of the *Supper* is almost entirely a monologue on the part of Theophilus, with little or no reference to the technical debate held during the supper itself. After the uncertainties of the final part of Dialogue 4, it starts off with, and largely consists of, a confident and composed restatement of Bruno's concept of universal infinity. Only at the end of this passage is it recognized once again that this new infinite cosmology is based on Copernican heliocentric astronomy. From the recognition that the earth moves, not just around itself with diurnal motion but also around the sun with annual motion, Bruno deduces the movements of all earths in the infinite universe around central suns which supply them with the necessary heat and light for life to exist. After Theophilus's definition of universal infinity in these terms, Smitho correctly points out the importance of the heliocentric proposal as the principal foundation of the Nolan cosmology. It is a theme that is too important to be treated as a mere digression. He requests that Theophilus end the cosmological part of his speech with a definition of the different movements of the earth, and Theophilus complies.

The premise, briefly outlined before the details of the earth's movements are defined, is closely related to Bruno's infinitism. The movements of the earth are first explained causally as necessary for the life-giving heat to penetrate the earth at as many points and in as many regions as possible. For this reason the movements of the earth will be many and complex. As can be seen, and has already been pointed out

49. *Atti della Reale Accademia Nazionale dei Lincei* (Rome) 7(1938–42):239, 244.

persuasively by Alfonso Ingegno in his book *Cosmologia e filosofia nel pensiero di Giordano Bruno,* Bruno's Copernicanism differs from the mathematical astronomy of Copernicus in that it is a life-orientated, biological cosmology that explains the movements of the earth in terms of the maximum life force it can obtain from the sun in its various movements around it.[50] Thus, although the diurnal movement around its own axis and the annual movement on an orbit around the sun are proposed in orthodox Copernican terms, Bruno is above all concerned to explain these movements in terms of a respiratory necessity of the earth, which causes the diurnal movement, and a reproductive necessity, which causes the annual movement. Although Bruno recognizes the importance of a mathematical determination of the earth's movements and of astronomical phenomena generally and often praises those who have assumed the task of supplying such mathematical determinations, he repeats that his own inquiry is not concerned with mathematical astronomy. Openly using the data supplied by others as its base, his inquiry is concerned with physical causes and effects; and it is in this field that Bruno claims to put forward valid and original ideas.

Bruno thus develops an entirely different treatment of Copernicus's third movement of the earth in one of his major innovations to the current cosmological debate. That he takes this approach is surprising as Copernicus's third movement was intimately linked with the phenomenon of the precession of the equinoxes, which Bruno in the later *De immenso* would claim to have been the subject that led him to turn to the *De revolutionibus* in the first place. Copernicus's third movement consists of a slipping back of the earth's axis during its annual revolution in a way designed to keep its direction constant with respect to the ecliptic (which in the Copernican astronomy is the name given to the path of the earth around the sun). This explanation is linked to an idea of bodies in movement around the sun attached to solid celestial revolving orbs. If the earth were carried around the sun on an orb, its axis would change direction at every point of the orbit and return to its original direction only when it regained its point of departure. Bruno, however, had already claimed that solid celestial orbs were pure neo-Aristotelian fantasy and had eliminated them from his cosmological discourse. Already the Nolan, during the supper, had put forward the idea of bodies suspended freely in an infinite ethereal space, without the cumbersome orbs to determine the nature of their movements. Theophilus can thus ignore this aspect of

50. Ingegno 1978:63–70.

Copernicus's third movement, reducing it to a form of rest, and concentrate on that part of Copernicus's third movement which refers to the precession of the equinoxes.

Having defined his third movement in terms that allow the axis of the ecliptic to remain constant, Copernicus then explains that this constancy is not absolute. The direction of the axis slips back very slowly over the centuries with respect to the sphere of the fixed stars, giving rise to the phenomenon known as the precession of the equinoxes.[51] Copernicus points out that to explain this phenomenon the traditional astronomy had been obliged over the years to introduce a ninth and even a tenth sphere beyond that of the fixed stars because it explained the phenomenon of precession in terms of the slow movement of the sphere of the fixed stars. This explanation appeared cumbersome, however, and had caused much unease because the extra spheres were visualized as empty of stars themselves: not a solution which Aristotelian doctrine readily allowed. By explaining the precession as a very slight movement of the axis of the earth in its orbit around the sun, Copernicus had much simplified the phenomenon of precession, and he was quick to point this out as a distinct advantage of his system over the traditional geocentric astronomy. Bruno's third movement of the earth in *The Ash Wednesday Supper* clearly refers to the precession of the equinoxes in Copernican terms. Its cause is said to be the necessity of an entire renovation of the earth's surface "over the centuries"; and the duration of an entire cycle, from the position of the zodiac overhead at one vernal equinox until the return of the same star pattern at another vernal equinox, during which time the northern hemisphere takes the place of the southern hemisphere to return at last to its original position, is given as 49,000 years. Curiously, Bruno is making a calculation here on the basis of the Alfonsine tables instead of giving the Copernican estimate of the full cycle of precession as 25,816 Egyptian years (that is, years of exactly 365 days), which is very nearly correct.[52]

Bruno then goes on, in the final pages of the *Supper*, to define a fourth movement of the earth, necessary, he claims, to guarantee mutation in the expression and complexion of the earth's surface. This phenomenon consists of a slight irregularity in the motion of the apsidal line of the earth. Many commentators claim that Bruno's fourth movement is en-

51. See Swerdlow 1975:49–98 for Copernicus's theory of precession.

52. In the *Spaccio* (see Bruno 1958:577) Bruno reduces his figure to 36,000 years, which was the figure given by Albert of Saxony. See *In libros de Caelo et Mundo*, lib. II, quaest. VI.

tirely autonomous because Copernicus himself recognizes only three movements of the earth. This claim, however, is erroneous. Copernicus includes several anomalies of precession in his account of the third movement; and under his fourth heading Bruno considers only one of the anomalies considered by Copernicus under his third movement.[53]

Copernicus, in book III, chapter 3, of *De revolutionibus*, explains the irregularities of precession, from a heliocentric point of view, as anomalies that have to be added to the varying obliquity of the earth's axis in order to save the phenomena.[54] One of these anomalies can be visualized as a very slight rotatory movement of the earth's poles about the zodiacal poles, varying the obliquity even further with a nonuniform motion, a phenomenon Copernicus describes as an *intorcta corolla* and which is usually known as axial precession. Copernicus calculates it as taking 3,434 years to complete. The other major anomaly is an oscillation backward and forward on the circle of precession, which Copernicus (on what seem to have been rather arbitrary principles) calculates to take exactly half the period of the variation of the obliquity, that is 1,717 years, to complete.[55] To accommodate these and other anomalies, Copernicus gives a table of calculated "corrections," which, together with the phenomenon of precession, should give the position of the equinox and the obliquity of the ecliptic at any given time.

Bruno, for his part, defines his fourth earth movement rather clumsily as the movement by which "the orientation of the apse of the earth toward the point close to the Arctic changes, together with the same orientation of the other apse toward the opposite point close to the Antarctic pole" (*"un altro moto, per il quale l'abitudine, ch'ha questo vertice de la terra verso il punto circa l'Artico, si cangia con l'abitudine, ch'ha quel-l'altro verso l'opposito punto de l'Antartico polo"*). Gosselin and Lerner are, in my opinion, mistaken in translating Bruno's "si cangia con" in this

53. I am contradicting here the account of the fourth movement of Bruno's earth given by Ingegno 1978:67, who has, in my opinion, been led astray on this point by Schiaparelli and Tocco, cited as his sources. The most faithful and sympathetic account to date of Bruno's four earth movements is given by Vedrine 1967:228–29.

54. The most complete comment on Copernicus's own account of the various anomalies involved in precession is in Swerdlow and Neugebauer 1984, on Book III of Copernicus's text. At page 128 Swerdlow points out the complications of Copernicus's treatment of this problem, identifying five "secondary motions" with respect to precession. Of the five, however, Bruno seems to be concerned only with the nonuniform motion of the apsidal line of the earth.

55. For the details of Copernicus's complicated calculatory processes, see ibid. Swerdlow and Neugebauer remark: "Anyone who thinks that Copernican theory is 'simpler' than Ptolemaic theory has never looked at Book III of *De revolutionibus*" (127).

passage as "exchanges with."[56] Bruno had already described the changing over of the two hemispheres through the phenomenon of precession as his third motion. The fourth motion, or axial precession, is a kind of movement of eight of the earth's apse around the arctic pole, "together with" an accompanying movement of eight of the opposite apse around the antarctic pole. However, it must be admitted that Bruno fails to express himself clearly here, and Giovanni Schiaparelli, in the nineteenth century, also read this passage as meaning an "exchange" of poles.[57] Bruno's definition, however, appears to correspond correctly to the anomaly known as axial precession, or the nonuniform motion of the apsidal line of the earth, considered by Copernicus as part of his third movement, although he may have confused it with the anomaly of the oscillation on the line of precession, or the variation in the rate of precession, as well as the other anomalies considered by Copernicus. In any case, Bruno gives no figures for the period of axial precession. Rather, he is concerned to point out that the overall movements of the earth are so complex as to make exact calculation almost impossible.[58] There is clearly some tension here in Bruno's mind, which will never be completely resolved. On one hand, exact astronomical calculation appears to be seen as a kind of net thrown over and capturing the life-giving qualities of the universe, an idea derived from an earlier Hermetic phase. Bruno claims more than once that all attempts to reduce the earth's movements to order are destined to fail because no one movement can be reduced to a regular geometrical line. On the other hand, he will recognize that the earth's movements do possess "a certain order" that must be considered as susceptible of more or less precise measurement.

To sum up the Copernican pages of *The Ash Wednesday Supper* is no easy task. Bruno clearly attempted to provide the sixteenth-century reader not with a series of technical footnotes to the *De revolutionibus* but with an original, philosophical reading and critique. He immediately distinguishes that reading from the prevalent interpretation of his period, particularly associated with scholars from Wittenberg, which stressed the mathematical virtues of Copernicus's astronomy and ignored its cosmo-

56. Bruno 1958:164, Bruno 1977:220.

57. An interpretation that was mistakenly followed even by Vedrine 1967:228.

58. Bruno has some justification for this position. Considering the overall model of Copernicus's calculations of precession, Swerdlow and Neugebauer 1984:73 write: "The principal source of difficulties throughout the precession and solar theories of Book III was the variation of parameters over long periods. This would be difficult enough if the parameters were correct, but mostly they were not, resulting as they did from very sensitive derivations from less than accurate observations."

logical implications. Perhaps overreacting to this approach, Bruno explicitly claims to see in Copernicus's book a physical cosmology rather than a mathematical astronomy, and his main intention is to draw out all the philosophical implications of that cosmology, as he sees them, with very few concessions to the theological and neo-Aristotelian objections it was raising. Although Bruno's approach gives rise to occasional mistakes in reading, in particular with respect to the question of lunar theory, he shows an overall understanding of Copernicus's text which is much more sophisticated than usually supposed and includes some of the most technically difficult aspects of the new astronomy such as the precession of the equinoxes and even anomalies such as axial precession. Furthermore, Bruno reaches daring and original conclusions that go well beyond anything included in, or authorized by, Copernicus's text. Most of these conclusions would become part of the astronomical debate of the following century.

Bruno was among the first to eliminate from the sky the celestial orbs and to conceive of all the celestial bodies, including earth, as freely suspended in a homogeneous ethereal space. Furthermore, he opened up the universe to a multiplicity of solar systems, thus foreseeing the intergalactic universe. Above all, his daring and uncompromising development of universal infinity launched into the post-Copernican cosmological debate a discussion which, although in rather different technical terms, is still ongoing. To stress that none of these achievements led to the technical progress made by great mathematical and observational astronomers such as Tycho Brahe, Kepler, and Galileo only a few decades later is beside the point, particularly when added to the assumption that Bruno's Copernicanism was limited to a mystical intuition of the sun's centrality, or of the infinity of the whole. Bruno's approach was different both from that of the mathematical astronomers and from the neo-Platonizing Hermeticists. It was an attempt to construct a philosophical basis not only for a new astronomy but for a new science.

Bruno's awareness of the scientific revolution which was developing in the wake of the Copernican destruction of the traditional cosmology was complex and far-reaching. He understood that the new inquiries would involve radical changes in the social and political structures of society, in the traditional doctrines of knowledge, in the relations between natural philosophy and theology, and in the logic of scientific discovery. The works that follow *The Ash Wednesday Supper* are concerned with issues such as these rather than with furthering his reading of Copernicus. The *Supper* thus presents Bruno's reading of Copernicus both in the form in which it matured after a long and complex process of cosmological meditation and as a starting point for a wider and more thorough investigation of the philosophical implications of the new cosmology. Only at the end of that wider philosophical investigation did Bruno return, in his Latin masterpiece *De immenso,* to a detailed analysis of his reading of Copernicus's cosmology.

4 Beyond Copernicus: *De immenso et innumerabilis*

The eight books of the *De immenso*, the final work of Bruno's Frankfurt trilogy published in 1591, are all concerned with the principal cosmological issues raised in his previous works, above all in *The Ash Wednesday Supper* and the *De l'infinito* of 1584, as well as the transitional *Camoeracensis acrotismus*, the account of Bruno's Parisian anti-Aristotelian debate of 1586 published at Wittenberg in 1588. The books of the *De immenso* define the nature and infinity of space; discuss the Aristotelian arguments in the *Physics* and the *De caelo* against an infinite universe; redefine the movements of the earth as part of one solar system within an infinite number of worlds; discuss the matter of which all earths and all suns are composed; deny the Platonic and neo-Platonic transcendental ideas; and equate the divine light with the light that illuminates an infinite and divine universe. Because Bruno continues to maintain that this philosophical concept of the immense universe is founded on an idea of heliocentricity as the defining characteristic of each of the infinite number of solar systems, he continues to consider his reading of Copernicus as one of the foundation stones of his infinite cosmology. By now, however, he has integrated Copernicus in a discourse that begins with an argument already fully developed in antiquity; and Bruno insists much more strongly here on the classical precedents for both his heliocentricity and his infinitism than he had in *The Ash Wednesday Supper*. Pythagoras is still frequently mentioned. Moreover, the very fact that the *De immenso* is a scientific poem in Latin, clearly modeled on the *De rerum natura* of Lucretius, underlines the Epicurean precedent for cosmological infinity.[1]

Bruno refers to Copernicus only in the third book of his text, and then in the final ninth and tenth chapters. By that time, he has already illustrated all the major implications of heliocentricity as he understood it, and Copernicus appears more as a voice offering technical confirmation of conclusions already reached by Bruno independently than as the heroic pioneer which he had seemed in the earlier Italian dialogues. This is the strategy Bruno follows in the brief autobiographical passage in book III, chapter 10, of the *De immenso* delineating his personal discovery and reading of the *De revolutionibus*. It is also the strategy that informs his wider reference to Copernicus in this final cosmological work.

Bruno is once again generous in his recognition of the Polish astronomer's achievement. Chapter 9 of book III is titled "The Light of Copernicus," and the image of illumination continues to dominate the discovery of the new cosmology. Whereas in *The Ash Wednesday Supper*, Bruno had quoted a large part of the preliminary letter of Andreas Osiander in order to ridicule it and expose its deceits, here he quotes and then paraphrases the greater part of Copernicus's own introduction to his text in order to praise it for its audacity and clarity.[2] The rest of the ninth chapter of book III defines the three movements of the earth contemplated by Copernicus. These pages are very carefully set out and exact. They appear almost to assume a didactic form and to be defined by an anxiety to get the technical details right. They are illustrated by figures copied directly from Copernicus, and the text is also a straight synthesis of the Copernican arguments for the three movements. A short paragraph at the end adds Copernicus's own comment on the question of precession, underlining the more convenient solution of an earth-oriented precession with respect to the addition of a ninth and tenth sphere. There is, however, no mention of the anomaly of axial precession, which, in *The Ash Wednesday Supper*, had constituted Bruno's fourth movement of the earth, or of the oscillation on the line of precession traditionally known as trepidation.

In chapter 10, Bruno develops his critical comment of Copernicus's universe and then his own final cosmological proposition. This chapter con-

1. On Bruno's debt to Epicurus and Lucretius, see Horowitz 1952: chap. 2 and above all Monti in Bruno 1980 and Monti 1994.

2. Bruno seems to have shared Westman's assessment of the importance of Copernicus's introductory letter to Pope Paul III (Lindberg and Westman, eds., 1990: 167–205).

tains some very unorthodox solutions that had not been proposed in the earlier and more fragmented discussion in *The Ash Wednesday Supper*. The first of these concerns the movement of the sun which, in the Italian dialogue, had been suggested but left undefined. Here Bruno proposes that the sun, which is at the center, not only revolves with a spiraling motion around its own axis but that it also moves around the center on an oblique orbit with respect to the celestial poles: the extremities of the orbit coincide with lines drawn parallel to the earth's two tropics.[3] Here he provides a solution in which the obliquity of the ecliptic (defined in the Copernican astronomy as the path of the earth around the sun) is eliminated to be replaced by the obliquity of a small solar ecliptic at the center of the planetary system. The advantages of this system, according to Bruno, are that it eliminates the problem of the third movement of the earth, and with it the necessity of imagining celestial orbs. Precession is defined in this system as a very slightly increasing inclination of the earth's axis from a position parallel to the celestial poles.

This rather surprising final cosmological proposal is not easy to visualize, particularly as no diagram is provided. Help in this task, however, can again be gleaned from Gilbert, whose second scheme of the earth-sun relationship attributed in the *De mundo* to "the Nolan" corresponds exactly to this page of the *De immenso*. Again it is worth quoting Gilbert's text in full. It is developed around Figure 6.

> Another way, according to the Nolan: Let DFCG be the colure of the solstitial points, in which C and D are the poles of the universe. Let AB be the equator of the universe around which the center of the earth A is continually moving from west to east. Meanwhile let the globe of the earth move around its center A on the axis of its poles c and d which are parallel to the poles of the universe, so that the earth's equator ab is in the same plane as the equator of the universe AB, and let it similarly move from west to east so as to create daily motion. Third let the sun describe around the same center E a small circle that is oblique to the equator. Let the common section of this circle with the colure mentioned above be gh and let its poles be G

3. Monti in Bruno 1980:610, n. 19, claims that Bruno was the first to consider the possibility of a movement of the sun around its own axis, which would become an important theme after Galileo's sighting of sunspots. Copernicus finds it necessary to make the sun move around a very small orbit at the center in order to save all the phenomena; and this aspect of his theory is correctly illustrated in the diagram of the universe published in first editions of his book. For Copernicus's solution to the movement of the sun, see Swerdlow and Neugebauer 1984.

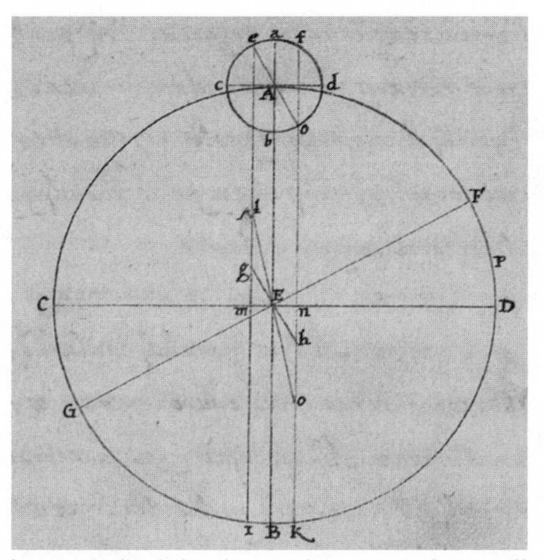

6. Diagram in the section headed "Alius modus iuxta Nol" in William Gilbert, *De mundo nostro sublumari philosophia nova*. This diagram is in the manuscript copy presented to Prince Henry and held by the British Library, call mark: ROYAL 12 F XI, f. 87v. By permission of the British Library.

> and F. Now draw the lines egi and fhk parallel to the equator. It is evident that when the sun is in h, it will determine one tropical point, and when in g the other, and when in E the equator; and sometimes the sun will be further away from the earth and sometimes it will come nearer to it. On this assumption it is clear that the greater the circle of the sun, the smaller the value one has to assign to the angle gEA, as is the case with the circle io [lo] whose pole is p [P], for the distance of the tropical points has to be maintained. It is also clear that the circle of the sun need not necessarily be oblique to the equator (unless it happened to have to be so for other reasons). Indeed, if the diameter of its circle were in m and n, everything would be the same. Third it is clear that hg or io [lo] do not represent the zodiac for us, but eo and ei [the point i is not given in the diagram] parallel to them on the great circle.[4]

Gilbert has confined his reconstruction of Bruno's final cosmological model to the earth-sun relationship; although Bruno himself goes on to develop another rather strange theory of the movements of Mercury and

4. Gilbert 1651:201; my translation.

Venus.[5] He sees them as orbiting around the sun on the same orbit as the earth, although on opposite sides of the orbit at any given time. It appears that Mercury occupies a central position similar to that of the earth on the other side of the orbit, while Venus becomes a satellite of Mercury orbiting around her on an epicycle, just as the moon moves on an epicycle around the earth. Bruno illustrates this aspect of his theory with the diagram already reproduced in Chapter 3 (Fig. 5). No explanation is offered as to how Mercury and Venus, according to this curious astronomical model, are supposed to maintain the known (differing) periods of their orbits around the sun which, in neither case, correspond to the solar year of the earth-moon orbit. Copernicus, in book I of *De revolutionibus*, gives Venus's period as nine months and Mercury's as eighty days. It is only fair to notice that Bruno admits to some confusion on this point. Shortly afterward, he seems to doubt his own solution, suggesting that earth, moon, Venus, and Mercury might all be circling on the circumferences of their epicycles: a suggestion that harks back to the Copernican discussion in the *Supper*. No calculations are given to support these cosmological hypotheses; and Bruno finishes the tenth chapter of book III of the *De immenso* by admitting that only professional astronomers have the technical capacity to prove or disprove them. Gilbert, as far as I know, was the only one of his contemporaries to take up Bruno's final cosmological scheme, and then only the part relating to the earth and the sun. Gilbert, himself no mathematical astronomer, also offers no calculations either to support or reject it.[6]

What Bruno is trying to do in the *De immenso* is clearly different from what he was doing in *The Ash Wednesday Supper*. There the generous praise of Copernicus was linked to an attempt to work out the details of the Copernican universe, though not without difficulties and some confusion over lunar theory. At the same time the Nolan philosophy was presented as an extension to infinity of the conceptual implications of the new heliocentricity, not fully realized by Copernicus. In the *De immenso* the technical details of the Copernican scheme are shown to have been acquired, at least as regards the essential question of the movements of the earth. Then that scheme itself is questioned and an alternative heliocentric model submitted. Bruno recognizes that he lacks the technical

5. Not much attention has been given to Bruno's final cosmological proposal. The most extended comments are in Michel 1962/1973:221 and McMullin 1987:59–60. Westman 1977 noted that it was rather "extraordinary." See Bruno 1879–91, I, i:395–98.

6. Gilbert's lack of mathematics was publicly admitted by one of his admirers, a student of the magnetic philosophy. See Ridley 1617:9–10.

capacity to back up his alternative solar system with observational and mathematical data. In that case, why did he propose it? And why was Gilbert interested enough to reproduce it?

After the brief page in which he presents it, Bruno makes no further mention of his alternative heliocentric system in the final five books of the *De immenso*, which suggests that he had no illusions that he was himself solving the problem of the movements of the earth with respect to the moon, the sun, and the other planets in our own solar system. It must be remembered, however, that Bruno believed the universe contained an infinite number of solar systems, and there was every reason to suppose that one of them might correspond to the model proposed. Furthermore, Bruno also appears to want to suggest that Copernicus had not found the final solution to planetary movement either. Bruno understood more fully than most of his contemporaries the truly revolutionary impact of Copernicus's book on the culture of his time. Yet from the earliest pages of *The Ash Wednesday Supper*, he had been repeating that there were elements in the new cosmology which were fundamentally erroneous and misleading. He centered his criticism on Copernicus's mathematical methodology and his lack of physical reasoning, because he thought that Copernicus was confusing mathematical concepts and physical realities. In this respect, Bruno felt that there had been little improvement on Ptolemy. The sky in which thinking people lived at the end of the sixteenth century was still cluttered with eccentrics, epicycles, celestial orbs, and precessional anomalies which were clearly conceptual tools interfering, in Bruno's opinion, with a visualization of the real shape of the cosmos. Bruno wanted these concepts, which Copernicus had inherited from the traditional cosmology, to be recognized as purely mental tools. They should be flexible where they had to be used, and where possible be eliminated altogether. It would, of course, be Kepler, with the discovery of the elliptical orbit of Mars, who would initiate the breakthrough in this context; but by that time Bruno was dead.

From Bruno's sixteenth-century perspective, it was essential above all to eliminate the celestial orbs, for it was their outer shell that defined a closed universe. This elimination can be seen as the basis of Bruno's alternative proposal, for it presented him with a problem. The traditional astronomy had satisfactorily explained the regular motions of celestial bodies by placing them on mutually touching circular orbs put in motion by the prime mover of Aristotelian fame. To eliminate the orbs meant eliminating the cause of the movements of the worlds in space. Bruno's realism in cosmology made him acutely aware of this problem. In *The Ash Wednesday Supper*, as we have seen, he attempts to explain celestial motions in biological terms as the "necessity" for the earths to expose themselves to the heat of the suns in varying times and with varying and complex movements. He thinks of every world as imbued with an internal motor, or intelligence, which dictates its movements according to its need to preserve its life: a stimulus similar to that which spurs the spider to weave its web. Bruno clearly realized, however, that such explanations were difficult to quantify. Furthermore, they hardly explained the regularity of the movements in question. Sometimes Bruno fights off this objection by claiming that astronomical measurement is pure illusion: an attempt to impose regularity on a fundamentally chaotic reality. But then he will concede that there is a basic regularity to the movements of celestial bodies, which can be measured at least in approximate terms. His problem then is to define why these regular motions occur, in the absence of the law of universal gravity later supplied by Newton. In the *Supper*, Bruno suggested Telesius's thermodynamic theory as an explanation of the movements of all the bodies in his infinite universe. In his final cosmological solution in the *De immenso*, Bruno is clearly abandoning this explanation for a more mechanical one. Mercury and Venus are placed on the opposite side of the orbit of the earth and the moon as a "counterbalance," in almost Archimedean terms. The model reminds one of the many machines of perpetual motion that were being constructed at this time.[7]

It may well have been this attempt to explain the movements of celestial bodies, in opposition to the mathematical astronomy of a Copernicus still dependent on a prime mover, that made Bruno insist on the importance of his model in spite of its technical deficiencies. Gilbert's interest may have been stimulated for the same reason. But it is probably no coincidence that Gilbert fails to reproduce the planetary aspect of Bruno's final model, limiting himself to the aspects concerning the sun and the earth. For Gilbert himself was not interested in a purely mechanical explanation of the movements of celestial bodies. The cosmological aspect of his work on magnetism can rather be seen as an attempt to integrate Bruno's cosmic intelligences with the idea of magnetic force as the explanation of the movements of the worlds in space.

7. One such machine, which was presented to James I, is described by Thomas Tymme in his *Dialogue Philosophicall* published in London in 1612. Tymme, who was also interested in alchemy and is known to have prepared a translation that has not survived of John Dee's *Monas Hierogliphica*, precedes his description of the machine with a cautious and largely disapproving discussion of the new cosmology which nevertheless shows some knowledge of the terms of the post-Copernican debate. Bruno is not specifically mentioned in this dialogue, although some of his cosmological ideas are hinted at.

Gilbert was most concerned to use his magnetic principle to explain the movement of the moon around the earth, as well as of a freely hanging earth, no longer encumbered by attachment to a solid orb, around a central sun. His magnetic explanation allowed him to resolve the problem of lunar theory as Bruno had not; the smaller body becomes for Gilbert necessarily a satellite attracted into the orbit of the larger one. Even though Bruno himself never arrived at a satisfactory explanation of the movement of the moon, Gilbert's extended references to his cosmological theories as they had been presented in the *De immenso* clearly suggest that reading him had been important. Like Gilbert, Bruno had put at the center of his cosmological speculation the problem of the cause of the movements of the worlds in a space now free of the cumbersome celestial orbs. To both Bruno and Gilbert, the answer to this question appeared of primary importance with respect to mathematical calculation. Gilbert came up with the better answer, as Galileo would recognize later on in his own cosmological dialogue.[8] The *De mundo* clearly suggests, however, that it had been important for Gilbert to be able to refer to the work of someone he believed had been asking the right questions.

8. See day 4 of the *Dialogo dei due massimi sistemi del mondo* in Galilei 1890–1909, vol. 7.

5 Bruno and the Gilbert Circle

William Gilbert was born in 1544 and died in 1603 of the plague, after a successful medical career that included the presidency of the Royal College of Physicians and culminated in his appointment as physician-in-ordinary to Queen Elizabeth's court. Gilbert was thus almost an exact contemporary of Bruno, who was born in 1548 and was burned at the stake in 1600. Gilbert studied at St. John's College, Cambridge, where he eventually became a senior fellow. He appears to have left Cambridge in 1581 and to have taken up residence in London. He would thus have been in London during Bruno's visit between 1583 and 1585. Gilbert seems to have conducted experiments in magnetism from the 1570s onward; for Edward Wright's preface to the *De magnete* claims that the work was ready for publication in the early 1580s.[1]

Gilbert apparently wrote the *De mundo* after completing the *De magnete*. It is made up of two parts, "Meteorologia" and "Physiologia." Scholars of this text have concluded that "Meteorologia" (the second part of the published book) was probably started in the 1580s and left incomplete, while the "Physiologia" was started in the 1590s and was still being worked on by Gilbert at his death. Whereas the *De magnete* consists of a soundly researched discussion of the magnet, promising practical advantages particularly in the field of navigation, the *De mundo* attempts a far more ambitious discussion of the philosophical and cosmological impli-

1. ". . . that work held back not for nine years only, according to Horace's Counsel, but for almost other nine . . ." See the "Address by Edward Wright" in Gilbert 1600/1958:xliv.

cations of magnetism. Recent studies of Gilbert have been characterized by the attention paid to the *De mundo* which the previous critical tradition had tended to treat as a marginal and unsatisfactory work.[2]

Contemporary evidence highlights the fact that Gilbert himself had a limited mathematical and astronomical training. In London he was part of a group including Edward Wright, author of the "Encomiastic Preface" to the *De magnete*, Mark Ridley, author of *A Short Treatise of Magnetical Bodies* (1613), and William Barlowe, author of *Magneticall Advertisements* (1617), which caused disagreement with Ridley over the association of the Copernican cosmology with the magnetic philosophy. Other members of the group were Launcelot Browne, a physician and Gilbert's closest colleague, Thomas Blundeville, and Professor Henry Briggs of the recently established Gresham College who worked with Gilbert on determination of latitude. Correspondence among these men establishes that Gilbert himself was not to be considered "skilfull in Copernicus" and that the astronomical and cosmological parts of the *De magnete* were heavily dependent on outside aid. In particular mention is made of Dr. Joseph Jessop, fellow of King's College, Cambridge, and later a fellow of the Royal College of Physicians presided over by Gilbert. Jessop is said by Ridley to have been "much esteemed" by Gilbert and to have lodged in his house in London. Ridley describes Jessop as "a great scholler in the Mathematick" and clearly suggests that his hand is to be detected in the more technical and astronomical parts of the *De magnete*, particularly in the sixth book. It should therefore be borne in mind that Gilbert may have collaborated with Jessop in preparing the cosmological pages of the *De mundo*.[3]

Students of Gilbert's theory of matter and cosmology have suggested that, far from limiting his work to a scientific study of the magnet, Gilbert was engaged in a project to develop a philosophy of magnetism as the unifying principle of the cosmos. The project is intimately connected with the new heliocentric cosmology, although Gilbert's acceptance of the Copernican universe was cautious and limited to those phenomena which lent themselves to explanation in magnetical terms. The phenomenon that most interested Gilbert was the movement of the earth, which

2. Boas 1951 explains the importance of the *De mundo* and stresses that a manuscript of this work was among Bacon's papers at his death. The first and only printed edition of *De mundo* was published from Bacon's manuscript. The first and so far only full-length study of this work is Kelly 1965. Stephen Pumphrey, in his Ph.D. thesis on Gilbert's magnetic philosophy, also dedicated considerable attention to *De mundo*.

3. For these remarks about Gilbert and Jessop, see Ridley 1617:9–10.

he considered in the *De magnete* only in its diurnal motion.[4] The later *De mundo* clearly suggests that this should not be interpreted as a cosmological statement limiting the earth's movement to diurnal rotation. Rather, for the purposes of what he has to say about magnetism, he is commenting on the daily motion, and in the *De magnete* Gilbert declines to go into the question of other possible movements of the earth. His principal aim, like that of Bruno, is to eliminate the heavenly orbs, which he sees as crashing around the heavens with their enormous size and weight, while the same phenomenon of diurnal motion can be more readily and appropriately explained by the smooth motion of a light and agile earth. "The philosopher never can admit such enormous and monstrous celestial constructions" is Gilbert's final comment on the celestial spheres.[5]

With the elimination of the celestial orbs, Gilbert can, like Bruno, reduce Copernicus's third motion to a form of rest. For what Gilbert calls the "steadfastness" of the earth's poles is not caused, as Copernicus had maintained, by a slipping back of the earth's axis as it circles around on its celestial orb. Rather, the earth hangs freely in the ether: a suggestion already made by Bruno in *The Ash Wednesday Supper*. Gilbert, however, can explain the stillness of the earth with respect to its poles better than Bruno could. For Bruno's thermodynamic philosophy could only suggest in negative terms that there was no reason for the earth to change the direction of its axis. This explanation of the problem in Bruno, as in the later *Dialogue* of Galileo, could be taken as a prelude to a theory of inertia; although neither Bruno nor Galileo explicitly define such a theory. Gilbert, instead, could use his magnetic philosophy to claim that such stillness derived from the earth's own forces, according to the magnetic powers within it that keep the direction of the earth's axis constant with respect to its poles.

Having eliminated as monstrous chimera the celestial orbs, as well as the idea of a *primum mobile*, Gilbert proposes to account for earth motion in animistic terms. He often uses a traditional neo-Platonic vocabulary to describe this motion. In the sixth book of the *De magnete* he writes: "By the wonderful wisdom of the Creator, therefore, forces were implanted in the earth, forces primarily animate, to the end the globe might, with steadfastness, take direction, and that the poles might be op-

4. On Gilbert's theory of matter and cosmology, see Freudenthal 1983. This study is based on the *De magnete* only and has nothing to say about Gilbert's relationship to the work of Bruno.

5. Gilbert 1600/1958:353.

posite, so that on them, as at extremities of an axis, the movement of diurnal rotation might be performed."[6] Such an account of diurnal motion, however, fails to satisfy Gilbert completely. Magnetism can help to explain the mechanism of the diurnal motion of the earth by supplying the reason why the position of the poles remains ordered and constant. Gilbert, however, needs a further teleological explanation of why diurnal motion takes place at all. Like Bruno before him, he wants to establish an efficient cause for earth motion, now that the system has remained without an outer orb of fixed stars through which the *primum mobile* started off the celestial motions in the traditional philosophy. Although failing to name at this point either Telesius or Bruno, Gilbert adopts the same thermodynamic causal explanation for the diurnal motion of the earth as Bruno had proposed both in *The Ash Wednesday Supper* and in the *De immenso*: "In her rotation the earth bears on her own poles . . . by natural conformity and fitness, as also for the sake of a necessary good and avoidance of ill, but most of all because the effused spheres of solar influence and of solar light do impel."[7]

The close link which Gilbert proposes between a magnetic force that controls the diurnal rotation of the earth and a thermodynamic necessity seen as the cause of that motion is reestablished shortly afterward in a passage that appears even more dependent on Bruno:

> The earth therefore rotates, and by a certain law of necessity, and by an energy that is innate, manifest, conspicuous, revolves in a circle toward the sun; through this motion it shares in the solar energies and influences; and its verticity holds it in this motion lest it stray into every region of the sky. The sun (chief inciter of action in nature), as he causes the planets to advance in their courses, so, too, doth bring about this revolution of the globe by sending forth the energies of his spheres—his light being effused.[8]

Gilbert stresses the cosmological revolution of his times above all in these causal terms. It had been a "ridiculous supposition" impossible for a philosopher to accept, that all the fixed stars, the planets, and the higher heavens, had revolved simply for the sake of the earth. "The earth then revolves, and not the whole heavens; and this movement brings growth and decay, gives occasion for the generation of animated things,

6. Ibid., 328–29.
7. Ibid., 329–30.
8. Ibid., 333.

and arouses the internal heat to productiveness."[9] Sixteen years earlier, in London, Bruno had written of diurnal motion in *The Ash Wednesday Supper*: "First, then, for its life and that of the things it contains, so that it can breathe in and out with the daily heat and cold, light and darkness, the earth moves about its own center with a regular period of twenty-four hours, exposing its surface to the sun."[10]

In an interesting page of chapter 5 of his sixth book of *De magnete*, Gilbert attempts to justify this animistic concept of diurnal motion with reference to ancient sources. The sources that must be repudiated insofar as they were responsible for the traditional belief in the movement of the sphere of the fixed heavens, dependent on the *primum mobile*, are "the early Peripatetics" and "the common run of philosophasters." Instead Gilbert recalls with approval the Stoics, who attributed to the earth a soul, claiming "amid the derision of the learned" that earth is an animal, and those sources which declare the earth's soul to be "astral."[11] Gilbert refrains from explicating his choice of vocabulary; although Frances Yates would presumably—I think correctly—have seen a Hermetic as well as a neo-Platonic influence here. The idea of a universal animism, regardless of the Aristotelian distinctions between the different kinds of soul, is strong in the Hermetic texts and probably should be seen as the element of Hermeticism which was most fully absorbed into the post-Copernican cosmological discussion, especially by those like Bruno and Gilbert himself who were looking above all for causal explanations of bodies in space.

There can be no doubt that one of the great strengths of Gilbert's magnetic theory with respect to Bruno's primarily thermodynamic one lay in its solution of the problem of the motion of the moon. He makes much of the subject in chapter 6 of the sixth book of *De magnete*, claiming that a "magnetical compact" between earth and moon compels both forward in motions that are proportioned between one another. Gilbert does not seem very certain here if this proportion is mathematical (a concept which, as he is careful to specify, had already been put forward by Aristotle in *De caelo*, book II, chapter. 10), or whether it depends on some "affinity" between them. Gilbert inclines toward the latter solution, using terms such as the moon being "of the same kin" as the earth, or the two being held together by certain "ties."[12] William Whewell pointed out long

9. Ibid., 334.
10. Bruno 1958:163–64; my translation.
11. Gilbert 1600/1958:339.
12. Ibid., 345.

ago that these rather vague notions of how magnetic "virtues" might affect the revolutions of the worlds in space were replaced, in the later *De mundo*, by a more distinct statement of the attraction exerted by the earth on the moon in terms of their relative mass: "The force which emanates from the moon reaches to the earth, and, in like manner, the magnetic virtue of the earth pervades the region of the moon: both correspond and conspire by the joint action of both, according to a proportion and conformity of motions, but the earth has more effect in consequence of its superior mass."[13]

The last part of the sixth and final book of the *De magnete* is dedicated to the problem of the precession of the equinoxes: the problem that encouraged Bruno to turn to Copernicus for explanations in the first place. Gilbert offers a detailed account of precession in Copernican terms, as multiple movements of the earth's axis and not of the sphere of fixed stars. He is more detailed than Bruno in his accounts of the anomalies, giving a fuller description of both axial precession, or the so-called *intorcta corolla*, and the movement back and forth on the line of precession itself, which had replaced the traditional phenomenon of trepidation. But what is of particular interest, with respect to Bruno, is his explanation of why precession should take place at all. Bruno had claimed that precession was necessary for "renewal over the centuries," that is, for long-term change in the characteristics and complexions of the different regions of the earth. Gilbert too says of precession, "hence the ever new changes of the seasons; hence are regions more or less fruitful, more or less sterile." Equally interesting is the similar conclusion as to the ultimate impossibility of measuring the movements of the worlds in space because the complexities of the movements seem to exceed by far the possibility of reducing them to mere numbers. Gilbert clearly endorses this position, which had already been forcibly expressed by Bruno, where he surmises that no doubt "other anomalies also will appear some centuries hence," to conclude his work on a strongly skeptical note:

> we still lack many and more exact observations to fix anything positively as to the anomaly of the movement of the precessions, as also of the obliquity of the zodiac. For since the time when in various observations this anomaly was first noted, only one half of a period of obliquity has passed. Hence all

13. The reference and translation are given in P. Fleury Mottelay's edition, where Whewell's publication is indicated as the *History of the Inductive Sciences* (1859), 1:394. See Gilbert 1600/1958:346.

> these points touching the unequal movement of precession and obliquity are undecided and undefined, and so we cannot assign with certainty any natural causes for the motion.[14]

Gilbert's skepticism here places him closer to Bruno than to Galileo, who would refer to his magentic philosophy so admiringly a few years later in his *Dialogue Concerning the Two Major World Systems*; although Galileo, it must be remembered, also chided Gilbert severely for not supporting his observations with mathematical calculation.[15]

Having established the close relationship between Gilbert's explanations of the movements of the bodies in space and a number of the cosmological ideas proposed by Bruno, one may wonder whether Gilbert was involved in more than an interested reading of Bruno's texts. At his trial, Bruno explicitly declared that *The Ash Wednesday Supper* was based on discussions held in London with some "doctors"; and Giovanni Aquilecchia suggested in his 1955 edition of the *Supper* that Launcelot Browne, Gilbert's closest friend, might be the "doctor of your own name" who, as the messenger bearing the invitation informs the Nolan, was one of the supper guests.[16] As no one by the name of Browne is mentioned as one of the guests seated at the table once the Nolan's party arrives, it is interesting to speculate whether Nundinius or Torquatus, whose important professional positions and fine clothes and jewelry are so heavily emphasized, could have been the Browne mentioned earlier on. For although Gilbert himself was clearly sympathetic to the Copernican astronomy, as well as to some aspects of Bruno's reading of it, there were some decidedly anti-Copernican members of the Gilbert circle.

It is possible to consider a scenario in which all the main speakers in the *Supper*, apart from the Nolan himself, belong to the circle of Gilbert's magnetically minded friends. Launcelot Browne, for example, who studied at St. John's College, Cambridge, and became an M.D. in 1576, was admitted to the Royal College of Physicians in 1584, the same year as the composition of the *Supper*. His prestigious and remunerative new appointment might well have stimulated the arrogant pomposity satirized by Bruno in the figure of the richly robed Nundinius. Browne, in spite of his friendship with Gilbert, was undoubtedly a steadfast neo-Aristotelian, to the extent of calling his son Galen. His philosophical orthodoxy is

14. See Gilbert 1600/1958:348–58 for Gilbert's account of the precession of the equinoxes.

15. See Galilei 1890–1909, vol. 4.

16. See Bruno 1955:53.

underlined, as a positive characteristic, by an admiring protégé, Thomas Blundeville, who praises him lavishly in his rigorously Aristotelian-Ptolemaic cosmological work, *The Theoriques of the Seven Planets* (1602). Blundeville, who was also a member of the Gilbert circle, appears to have been the typical professional Elizabethan grammarian and pedantic man of letters. As well as a commentary on Plutarch, which he dedicated to the queen, he published in 1599 an *Art of Logike* of such arid neo-Aristotelian orthodoxy that it is tempting to see in him the model for Bruno's Prudentius.

As for the bigoted theologian, Torquatus, may he not have been William Barlowe, educated at Balliol College and therefore an appropriate candidate for one of those zealously Protestant opponents from Oxford with whom the Nolan was called upon to debate? Barlowe was the son of a noted reformed prelate, also William Barlowe, who expressed his intense religious zeal in didactic tracts. This legacy would make his son a good candidate for the Torquatus who tries to teach cosmology to the Nolan, earning for himself the sneering reply that even schoolchildren knew as much cosmology as he did.[17] In one of his angrier moments with Torquatus, Theophilus plays heavily on his sheeplike intellectual qualities, which seem to correspond well with Barlowe's name. The bitter introductory poem that prefaces *The Ash Wednesday Supper*, dedicated to "The Malcontent," was probably directed at whoever was the historical Torquatus, for again Bruno underlines the didactic reforming zeal of his opponent: probably the same person who later induced Fulke Greville, formerly a friend and protector, to move over into the camp of Bruno's enemies. In the poem, Bruno claims to have been savagely attacked by the "stick and sword" of the Malcontent, and he adds that if, in his reply, he should himself fall to the ground, the name of his enemy will be "in diamond writ." This is probably a reference to the traditional belief, which went back to Ptolemy, Plutarch, and Pliny and had been recently reaffirmed by the Neapolitan Battista Porta, that stroking a magnet with a diamond would cause it to lose its power.[18] Chronologically this so far unglossed line of the *Supper* of 1584 would make sense. Gilbert is

17. The *Dictionary of National Biography* describes William Barlowe the elder as a reformed prelate who managed to survive Henry VIII, Edward, and Mary to return to England under Elizabeth. He is known to have written numerous manuscript tracts characterized by their reforming zeal. They included an ABC for the clergy and a *Briefe Summe of Geography*.

18. "Some have thought that the magnet would leese this quality attractive, if he were annointed with garlicke, or touched with a Diamond, and of this opinion were Ptolemie, Plutarke, Plinie, and Baptista Porta": Ridley 1613:a1v.

thought to have studied the magnet since the early 1580s, and Barlowe claims that his own studies preceded those of Gilbert. Barlowe remained a faithful neo-Aristotelian to the bitter end, declaring in his *Magneticall Advertisements* his admiration of Gilbert as the father of the magnetic philosophy but refusing explicitly to follow him in the "absurd" post-Copernican cosmological developments of his thought. Barlowe wrote of the sixth book of Gilbert's *De magnete*: "But concerning his sixthe Booke entreating of the motion of the earth, I thinke there is no man living farther from beleeving it, than my selfe, being nothing at all persuaded thereunto, by the reasons of other men, which he alledgeth, and as little or lesse (if it were possible) by those his inventions, endeavouring to prove the motion of the earth by the earth's *Magneticall* force and virtue."[19]

Barlowe's *Advertisements* stimulated an angry reply from another magnetical philosopher, also a distinguished physician, Mark Ridley, author, in 1613, of *A Short Treatise of Magnetical Bodies*. This pamphlet testifies to Ridley's advanced Copernicanism, to the extent of contemplating with equanimity a universe in which all celestial bodies, and not just the earth and moon, are composed of the same "magneticall substance unknowne unto us."[20] There is no mention of Bruno. Ridley's universe, however, is clearly composed, like Bruno's, of a homogeneous substance throughout, with no distinction between the matter composing earthly and celestial bodies. Not many even in James's England were prepared publicly to endorse such a theologically loaded theory of matter, similar to the one put forward by Bruno himself. At a cosmological level, Ridley endorses Tycho Brahe rather than Bruno, even if Brahe himself refused to accept the idea of a homogeneous or "open" universe. Ridley, however, is clearly accepting a compromise solution that had found a considerable consensus in the culture of his time. It is tempting to identify Ridley with the open-minded but cautious Smitho, who converses civilly with Theophilus in the *Supper*, although claiming that it may be better for the philosopher to follow the common opinion, "so that in case of error he will not be without general approval and companionship."[21]

Ridley nevertheless makes very plain in his *Treatise*, and in his later public discussion with Barlowe, his contempt for the backward-looking neo-Aristotelianism, which refused to move into the larger spaces of the post-Copernican world. In his reply to Barlowe's *Advertisements*, pub-

19. Barlowe 1617:B1v.

20. Ridley 1613:154–55.

21. "*Smitho*. Però comunemente si va appresso al giudizio comone, a fin che, se si fa errore, quello non sarà senza gran favore e compagnia. *Teofilo*. Pensiero indegnissimo d'un uomo!" See Bruno 1958:48.

lished in 1618, he not only ridiculed Barlowe's refusal to take seriously the sixth, cosmological, book of Gilbert's *De Magnete*, but he wrote a page on the free movements of all the bodies in an ethereal space which reads remarkably like Bruno:

> Now I would learn what Ordinance, Law and Trade these Globes and Starres have, which are placed in this thin Firmament, which are coeternall with the world; and which shall continue as long as the world shall last, if it be not their indefatigable motion to move naturally, or erraticably like baly errants to turne about, to give and receive the lightsome beames and influences benefitiale one to another, and to make other motions about their Center, according unto the faculties given them at their creation.[22]

This passage, with its reference to the globes and stars as "baly errants," is particularly interesting in view of Bruno's illustration of the earth's various movements in the final pages of the *Supper* "as if it were a ball which, mounting up at first moves rapidly and then ever more slowly and returning downward inverts the process."[23] Furthermore, in a corollary to his main text, Ridley refuses the idea of a *primum mobile* as well as a sphere of fixed stars on the basis of precession. Instead, "God at the Creation gave virtue and power to the natural bodies of the starres to make their motions whether one or many, and not unto these fained Spheres." Diurnal motion and possibly the motion of the earth around the sun ("as some thinke") appear to be endorsed by Ridley, who nevertheless finishes off by invoking once again the more cautious cosmological solution of Tycho Brahe.

In his reply to Ridley's *Discovery* of his *Animadversions*, Barlowe assumes a righteous tone of rebuke which is remarkably reminiscent of Bruno's Torquatus: "Out of all question, somewhat it is more than ordinarie, that maketh him of so hauty a spirit, so to brave the world with such prodigous assertions of his Magneticals, in, and above the Moone; the earths Magneticall circular Motions, and his paltry abusing of the holy scriptures to support his lunaticke fictions under the name of Magneticall Philosophie." And with that mixture of English and Latin which is so reminiscent of Bruno's neo-Aristotelian pedants, Barlowe goes on to sneer at Ridley's "Magneticall Astronomy, with those topsiturvy motions, like *Volvitur Ixion, & se Sequiturque fugitque.*"[24] And if the accusation of madness, or "lunaticke fictions," related to the circular motions

22. See Ridley 1617:7.
23. Bruno 1958:167.
24. Barlowe 1618:8.

of heavenly bodies, is strikingly reminiscent of the accusation made by the Oxford dons that Bruno's "head did run round" when he attempted to "set on foote the opinion of Copernicus," even more remarkable is Barlowe's further castigation of Ridley's "unreasonable and senseless opinions, which have more neede of Helleborus to purge them out of his head, than arguments to confute them in his booke."[25] How is it possible to forget at this point those mutual accusations of madness between the Nolan and Torquatus in the fourth dialogue of *The Ash Wednesday Supper*, when "Doctor Torquatus, turning to his table-mates, brought forth from the profundity of the scabbard of his conceit, and mounted on the edge of his moustache, an Erasmian adage: *Anticyram navigat*?"[26] The Erasmian adage in question is a reminder of the classical island of the mad, Anticyram, to which those in need are sent to gather the herb hellebore, a powerful laxative supposed also to be a cure for madness. Theophilus remarks of Torquatus's reference to this adage: "I think he was foretelling (even if he did not understand his own prophecy) that the Nolan was going to provide hellebore, in order to glue together the brains of these mad barbarians."[27]

The almost uncanny way in which this later dispute between Barlowe and Ridley recalls the heated cosmological discussion in *The Ash Wednesday Supper* seems to suggest at least the possibility that the Gilbert circle, with their radical internal divisions on the question of cosmology, provided the context in which Bruno's cosmological speculation in London took place. The documentary evidence at present available is not sufficient to confirm this hypothesis. Gilbert's pages dedicated to "the Nolan" in the *De mundo* nevertheless demonstrate without any doubt his own knowledge of, and interest in, Bruno's later cosmological speculation as it was put forward in the Latin *De immenso*.

As a final comment on the Bruno-Gilbert relationship, I wish to draw attention to the illustration of Gilbert's own picture of the universe as it was given in the cosmological section of the *De mundo*. It may be no coincidence that this cosmological picture immediately follows the pages of comment on the two Nolan systems reproduced by Gilbert. His own universe has clearly, like Bruno's, been freed of the celestial orbs, and the earth, with its accompanying moon, circulate freely around the sun according to their own internal impetus. It has sometimes been claimed, particularly by Susan Kelly, that Gilbert never declared himself on the

25. Ibid., 7.
26. Bruno 1958:132; Bruno 1977:186.
27. Ibid.

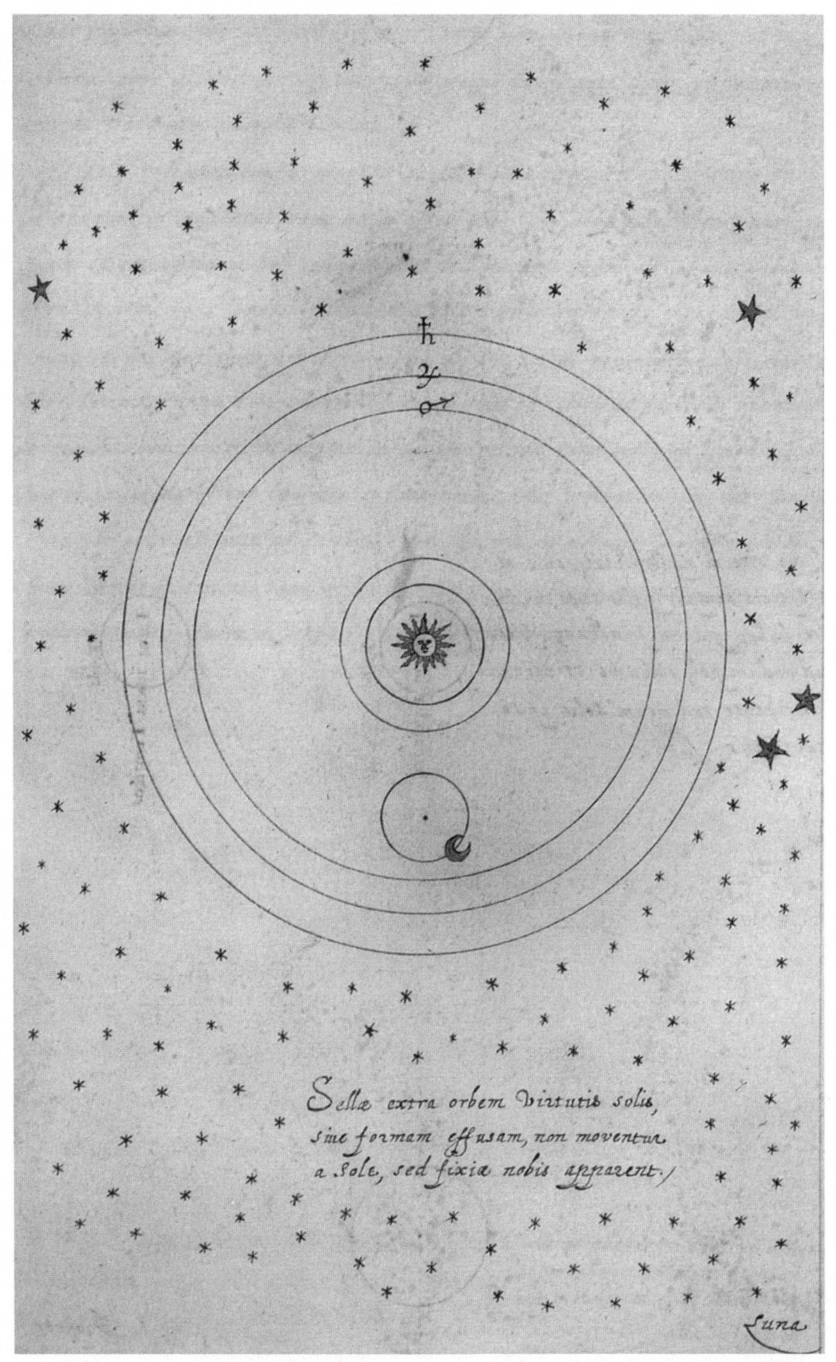

7. Diagram of the universe showing luminous bodies in gold leaf, in William Gilbert, *De mundo nostro sublunari philosophia nova*. This diagram is in the manuscript copy presented to Prince Henry and held by the British Library, call mark: ROYAL 12 F XI, f. 88. By permission of the British Library.

question of the earth's annual motion. Freudenthal, however, arguing on the basis of Gilbert's cosmological diagram, doubts the veracity of this claim. "Gilbert's diagram of planetary motion," he writes, "which avoids tracing any circle through the sun or the earth, in fact implicitly precludes the geocentric model. If the sun with the planets, as represented in the diagram, were taken to revolve around the earth, it would penetrate the zone of the fixed stars; that is, there would be an observable annual parallax."[28] Furthermore, outside the orbit of Saturn, Gilbert, like Thomas Digges before him, fills his sky with stars no longer fixed within an outer closing sphere. He appears to give no indication of how he thinks of this outer region of stars, all of them the same, it would seem, in printed versions of this text, which give this illustration in black and white. Gilbert's half-brother, however, presented to the young Prince Henry a handsome bound manuscript copy of Gilbert's text in which this illustration is delicately colored.[29] In it the sun at the center of our universe, together with the reflecting moon, is done in fine gold leaf; and in the same leaf some, but not all, the stars in the outer universe glitter like other suns. So it would seem that Gilbert is endorsing Bruno's idea of multiple solar systems in which ever more distant earths move around their central suns, even if Gilbert, like Galileo after him, refuses in the *De mundo* to make any explicit statement on the question of infinity.

28. Freudenthal 1983:35.
29. British Library, Royal and Kings Mss., 12.F.XI, f. 88.

6 The Infinite Universe

When Copernicus put the earth in movement, both around its own axis and around the sun, he radically changed the terms in which the mind contemplated universal space. From a solidly heavy and central earth, the traditional cosmology had seen a closed sky of stars and planets attached to massive celestial orbs revolving around it. Copernicus, on the other hand, accentuated the smallness and lightness of the earth, whose agile revolutions seemed to him so much more natural than the cumbersome revolutions of the enormous mechanism of solid orbs around a central earth. Although maintaining the concept of solid revolving orbs, Copernicus envisaged the outermost orb of the fixed stars as lying at a far greater distance from earth than the traditional cosmology had allowed. Without denying the idea of a closed world, Copernicus could thus talk of an "immense" universe that *appears* infinite and with respect to which the moving earth now seems like a tiny finite point. This new cosmological perspective is outlined by Copernicus in the sixth chapter of the first book of *De revolutionibus* titled "Of the immensity of the sky with respect to the size of the earth." It immediately led to a lively revival of a debate, begun in classical antiquity and continued with fervor throughout the Middle Ages, as to whether God could create or had created the universe of finite or infinite dimensions.[1]

The importance of Bruno's reading of Copernican heliocentricity within the post-Copernican cosmological discussion was once amply recognized

1. Duhem 1913–59 follows this discussion, claiming the continuity between medieval and early modern science. The classic discussion of the change from the idea of a closed to an open universe is in Koyré 1957.

by a long line of distinguished historians of science, from William Whewell to Alexandre Koyré, but recently it has been underestimated or ignored in the light of Yates's claim that for Bruno the Copernican diagram "is a hieroglyph, a Hermetic seal hiding potent divine mysteries of which he has penetrated the secret."[2] His extension of Copernican heliocentricity to a universe of infinite dimensions, on the other hand—the philosophical importance of which Dorothea Singer finely stressed in her translation of the *De l'infinito*—has remained at the center of much comment even in the post-Hermetic discussion of Bruno. For the Hermetic interpretation of Bruno's works writes off his Copernicanism as a "hieroglyph" but concedes ample attention to his infinite universe as a place of expanded, mystical gnosis: the gateway to reconciliation with the divine principle and cause. Such an interpretation of Bruno's infinitism can be accused of being off the mark. For if there is no doubt that it is through his concept of universal infinity that Bruno proposes reconciliation with the divine intelligence, it is less certain that Bruno's infinite universe was constructed entirely, or even primarily, on the basis of Hermetic sources. Robert Westman, in his reply to the Yatesian Hermetic interpretation of Bruno, demonstrated convincingly that the Hermeticists of the sixteenth century shared no common cosmological picture.[3] Moreover, the more recent comparisons of Bruno's cosmology with those Hermeticists who did extend the universe to infinity have all pointed out the differences between the Hermetic concept of universal infinity, which was essentially neo-Platonic with clearly theological implications, and Bruno's infinite physical space populated with an infinite number of worlds.[4]

Bruno's infinitism is one of the guiding axioms of his philosophy and is continually recalled throughout his philosophical works. It receives its definition and philosophical formulation above all in three major texts: the post-Copernican Italian dialogue *The Ash Wednesday Supper* (1584), the slightly later dialogue, also in Italian, *On the Infinite Universe and Worlds* (1584), and finally the Latin *De immenso* (1591). In all these works, the infinite universe is seen as an extension of the Copernican heliocentric cosmology, and therefore as intimately related to Bruno's reading of Copernicus himself. The relationship between Bruno's Copernicanism and his infinitism changes in the course of his philosophical development. In the Italian dialogues, Copernicus's heliocentric astronomy is praised before the Nolan proposes his vision of an infinite universe and

2. See Yates 1964:241.
3. See Westman 1977.
4. I shall be referring in particular to Granada 1992 and 1997, and Vasoli 1991.

appears clearly as its foundation and conceptual base. In the Latin *De immenso* ancient and modern pre-Copernican precursors of the infinite universe are given far more prominence; and the Copernican astronomy itself, although far from being disregarded, is encapsulated as a stage in the Nolan meditation on infinity rather than as its historical or conceptual basis. In spite of this later modification, however, Bruno's infinitism remains intimately embedded in his study of Copernicus. The new, astronomically defined heliocentricity, linked to his reading of Telesius, allowed Bruno to develop a concept of an infinite universe composed of an infinite, homogeneous substance in which cold earths revolve around hot suns to guarantee the process of universal life. This idea of cosmic infinity, involving an infinite number of solar systems, is essentially different both from that proposed by the ancients and from the Hermeticists, as well as from the theologically defined infinitism of neo-Platonic precursors such as Nicholas of Cusa.[5]

The first mention of the infinite universe in *The Ash Wednesday Supper* is found in Dialogue 1, immediately after the page of generic praise of Copernicus himself. It is couched in the form of a celebration of a new form of infinity which the Nolan philosophy proposes to elaborate on the basis of the newly heliocentric universe put forward in the *De revolutionibus*. Just as Bruno's first mention of the sun as center in his early *De umbris idearum* was proposed in decidedly Hermetic terms, without precise scientific connotations, so the first mention of infinity tends to suggest a Hermeticized intuition of an infinite region of the soul: the infinite space in which the soul can search for its reconciliation with the divine:

> We recognize many stars, many astral bodies, many protective spirits, which are those hundreds and thousands who accompany us in our praise and contemplation of the first, universal, infinite, and eternal efficient cause. Our reason is no longer imprisoned by the bands of the imaginary eight, nine, and ten revolving orbs. We understand that there is one sky, an immense ethereal region, where these magnificent lights keep their proper distances in order to participate in eternal life. These flaming bodies are ambassadors who announce the excellence of the glory and majesty of God. In that way, we are led to discover the infinite effect of the infinite cause, the true and living image of infinite power.[6]

5. The classic treatments of Bruno's infinitism are to be found in Koyré 1957 and Michel 1962/1973. Also important are Singer 1950; and Greenburg's introduction to his translation of *The Cause, Principle, and One* (Greenburg 1950).

6. See Bruno 1958:34, and Bruno 1977:90–91. The translation here is my own.

To understand Bruno's thought on universal infinity correctly, it is important to realize that within what appears at first sight, in this passage from *The Ash Wednesday Supper*, a purely mystical, even cabalistic, celebration of the infinite, Bruno inserts one phrase which indicates the more scientific train of thought he will develop in the following pages of his dialogue.[7] This is the reference to the "imaginary bands": that is, the external spheres of the fixed stars, and beyond, which closed the traditional universe within finite dimensions. Bruno uses the plural here because, as we have seen, the increasingly urgent problem of precession was being solved in his time, within the traditional Ptolemaic scheme, by the addition of a ninth and even a tenth sphere outside the eighth sphere of the fixed stars. These extra spheres, however, were also rigorously defined as closed and, in Bruno's opinion, only supplied extra conceptual prisons with which to suffocate the mind.

In his essay "The Greek Theory of the Infinite Universe," David Furley has shown that the ancient debate on universal infinity was characterized throughout by the argument as to whether something existed outside the "band" defining our own world.[8] One of the earliest and most colorful expressions of this argument was that of the Pythagorean Archytas, who, in the fourth century B.C., asked: "If I were at the edge of the world, as it might be in the region of the fixed stars, could I stretch out my hand or a stick into the outer region or not?"[9] Furley demonstrates how this idea of an "edge" of the world pervaded all ancient thought, even among those philosophers, like the Pythagoreans themselves, who put forward the idea that there *was* something outside the outermost sphere of stars.[10] Richard Sorabji, writing on the same subject, claims that the best answer to Archytas came from Alexander of Aphrodisias (in *Questiones* 3.12, 106, 35–107, 4), who argued for the impossibility of stretching the hand into nothing, an argument taken up by Bruno himself (although without mentioning Alexander).[11] Hippocrates' idea of a living universe envisaged the world as a "creature" enclosed within a living membrane or skin; while the Stoics thought of infinity as an empty space outside the boundaries of

7. For Bruno's use and reading of the Hebrew cabala, see De Léon-Jones 1997.

8. See Furley 1981.

9. Simplicius, *In Physica*, 467.26ff., quoted ibid., 578.

10. Furley 1981.

11. See Sorabji 1988, chap. 8: "Is there infinite or extracosmic space? Pythagoreans, Aristotelians and Stoics," and Bruno 1958:371, where the moderately neo-Aristotelian Burchio, in the *De l'infinito*, says: "se uno stendesse la mano oltre quel convesso . . . quella non verrebe essere in loco, e non sarebe in parte alcuna, e per consequenza non arebe l'essere."

the fixed stars.[12] Even the Democretean idea of the world as a microcosm, which was particularly congenial to the atomistic philosophy, saw our one world as a bounded version of the larger cosmic whole.[13] The atomic view was thus no exception, in Furley's opinion, to the terms in which the ancient debate on infinity was conducted. The Epicurean Lucretius used the image of our own world as a walled city *outside which* an infinite number of other worlds reeled around in a terrifying cosmic dance.[14]

The discussion of universal infinity in the Hermetic texts takes place within this classical context of thought. A quotation from the *Asclepius* which Yates used to define Bruno's idea of universal infinity makes this clear: "For as the space outside the world, if it exists (which I do not believe) must be, in my opinion full of intelligible beings, that is beings like the divinity of that space, so the sensible world is absolutely full of living beings."[15] Not only is Hermes here taking up a position which is explicitly traditional and conservative with respect to the radical infinitists such as the atomists, he is also positing an essential difference between the bodies that inhabit a possible space outside the world ("which I do not believe") and those in the sensible world. The "outside," for Hermes, is (possibly but improbably) "full of intelligible beings" who can be identified with the divinity of outer space. Our own world is full of sensible beings which Hermes, like Plato, sees as also alive and imbued with sparks of divinity, but, at the same time, as infinitely distant from the full intelligibility of divine being which begins beyond the fixed stars.

This concept, which is so essential to ancient discussions of universal infinity, is present even in the eloquent passage from another Hermetic text, the *Pimander*: the passage where the Magus man "leant across the armature of the spheres, having broken through their envelopes."[16] The concept here, which is essentially neo-Platonic, is that the magical doctrine of correspondences between the various grades of being allows the well-indoctrinated mind to cross the boundaries. Yates states that this is

12. See West 1971.

13. The atomists' view of infinity is clearly outlined, although in hostile terms, by Aristotle. See *Physics*, III, 4, 203b, 15–30.

14. Lucretius, *De rerum natura*, I, 1102; II, 1144; III, 16; V, 371, quoted in Furley 1981:574.

15. See Yates 1964:245 and Copenhaver, ed., 1992:87–88. Copenhaver's translation of this passage runs: "For just as that which is said to be 'beyond the world' (if there is any such thing, I do not believe that [it is void]) is full of intelligible things resembling it in divinity, as I take it, so also is this world that we call 'sensible' completely full of bodies and living things."

16. Quoted in Yates 1964:239.

what Bruno thinks too. She comments: "Bruno claims to be qualified as prophet and leader of the new movement because he has made an ascent through the spheres. Under the impression that the Copernican discovery has abolished the spheres to which the stars were formerly thought of as attached, he sees this as a breaking of those envelopes by which the Hermetic gnostic ascended and descended through the spheres."[17]

Our reading of Bruno's Copernicanism, however, has shown that Bruno never believed that the Copernican cosmology in its original formulation abolished the celestial spheres. He claimed that discovery for himself, arguing for it in precise terms, partly on the basis of precession, which, since Copernicus, could be explained as a slight movement of the earth's axis rather than as a lengthy revolution of a sphere of fixed stars, and partly, probably following Alhazen and also Jean Pena, on the basis of optical reasoning. The point about Bruno's reformulation of universal being in the wake of Copernican heliocentricity was that the ascent could now (theoretically) be a physical one into outer space, and not only of the mind, for there were no longer any envelopes to break through.

Bruno's infinite universe represents a conceptual rupture with the ancient discussion of universal infinity, including that in the *Asclepius*. The ancient discussion, as Furley has shown, was based on differing opinions as to what, if anything, lay "outside" the outermost sphere of the fixed stars. Bruno carries to its most radical conclusion his conviction of the essentially imaginary nature of that outer "band" to our own world. With the concept of bands, he eliminates any distinction between the kinds of being in the different parts of the infinite universe. This idea was accompanied by a frontal attack on Aristotle's doctrine of the hierarchies of the soul. In his answer to Nundinius's fourth proposition in *The Ash Wednesday Supper*, Theophilus claims in no uncertain terms that all the other globes in the universe "are earths, in no way different from our earth in species, but only greater or smaller." The movements of all these infinite bodies, continues Theophilus, are determined by the motions of their particular souls. When a shocked Nundinius asks him if he thinks of these souls as sensitive, Theophilus replies: "Not only sensitive, but intellective as well." At this point, narrates Theophilus, Nundinius fell silent and laughed no more.[18] What has happened here is that Bruno has floored Nundinius, who thought he was proposing, like Lucretius, a purely casual material universe that stretched out into infinity. On the contrary, Bruno

17. Ibid.
18. Bruno 1958:109–10.

is proposing a material universe imbued throughout with the order and intelligence of the divine mind.

This is quite a different idea from the infinitism of much Hermetic cosmology, as it is defined in the passage from the *Asclepius* quoted above. Scholars have demonstrated that the Hermetic infinity played an important role in sixteenth- and seventeenth-century cosmological discussion. It is becoming increasingly clear, however, that the Hermetic idea of universal infinity needs to be distinguished from the model put forward by Bruno. Relying still on the ancient concept of a band of fixed stars which might, or might not, have something outside it, the Hermetic texts allowed the development of compromise solutions which filled an infinite post-Copernican space with divine, intelligible beings, sometimes identified with the superior stars themselves. This was the solution proposed in England by the pro-Copernican Thomas Digges, whose infinite universe remained heterogenous insofar as a superior "outside," with respect to our own solar system, was seen as filled with stars imbued with the souls of angels and the elect.[19]

This peculiarly Protestant cosmological solution found a similar if not identical counterpart in southern Europe in the infinitism of the Italian Hermeticist Francesco Patrizi of Cherso. Patrizi's infinite universe has been the subject of an interesting and erudite comment by Cesare Vasoli, who emphasizes both the affinities and the differences of Bruno's and Patrizi's infinitism.[20] Taking as his central text the section of Patrizi's *Nova de universis philosophia* titled "Pancosmia," and in particular book VIII of that text, Vasoli underlines how Patrizi, like Bruno, was concerned to eliminate from a space that assumed infinite dimensions all the cumbersome physical and conceptual entities of the traditional cosmology such as the crystalline spheres with their necessary prime mover, together with a series of Peripatetic concepts such as *potentia*, privation or form.[21] Like Bruno, Patrizi considered his infinite universe a return to a more simple and natural habitat after the complicated mechanism of the "machina mundi" of the Aristotelian-Ptolemaic system inherited by Copernicus. Patrizi's universe, however, as Vasoli stresses, is based on the ancient distinction between a unique world and a surrounding, essentially different, "empyrean." Patrizi seems to have used the same ancient

19. For Digges's cosmology seen in its essential difference from that of Bruno, see Koyré 1957, Westman 1977, and Granada 1992 and 1997.

20. See Vasoli 1991.

21. There is no full-length English translation of Patrizi's text, but Brickman 1943 translates Patrizi's ideas about space.

sources as Gilbert: Stoic philosophy, with its idea of an infinite empty space beyond our own world, and the Hermetic texts which (insofar as they conceded physical infinity at all) filled up the extramundane space with divine light and life.[22]

Behind Patrizi's version of the infinite universe can be detected another Italian presence: that of Palingenius, author of the *Zodiacus vitae* published in Venice in 1531.[23] Palingenius's complex and often ambiguous text claimed for itself a Catholic orthodoxy that was belied by frequent strong attacks on the monastic culture of the time: an aspect of his work which gained him an enthusiastic public in northern Europe as well as an early translation into English.[24] Palingenius's work contains one of the most powerful renaissance reappraisals of Epicurean philosophy, although its Epicureanism, clearly derived from the poem of Lucretius, is curiously blended with a strong neo-Platonic and Hermetic strand of thought. Essentially an ethical work that underlines the good and bad characteristics associated with each sign of the zodiac, Palingenius's poem clearly wishes to establish a principle of pleasure and order within the natural universe. That principle, rather than being based on the casual agglomerations of Epicurean atoms, or even on the more pleasing cosmic whirlwinds of Democritus, calls rather upon a neo-Platonic concept of light: a natural light as the primal substance of a unique world and a divine light as the essence of the infinite space outside.

Bruno, like Patrizi, had read Palingenius with interest and admiration, although tempered with criticism that separates Bruno's infinitism both from that of Palingenius and of Patrizi. In chapter 6 of the eighth and last book of the *De immenso*, published in the same year as Patrizi's *De nova universis philosophia*, 1591, Bruno expresses his opinion of Palingenius in a chapter significantly titled "Palingenius dreams profoundly together with Plato when he thinks of infinite light deprived of body as lying beyond this finite world and beyond the visible stars."[25] Bruno claims here that his own concept of universal infinity is based on the idea of a continuum, which is a necessary logical conclusion of the idea of universal

22. The importance of Patrizi's ideas in sixteenth-century England, as well as those of Telesius and Bruno, has been underlined in Henry 1979 and Prins n.d.

23. For Bruno and Palingenius, see Granada 1992.

24. See Palingenius 1531 and 1565/1947.

25. *De immenso*, bk. VIII, chap. 4. As well as Platonic and Hermetic sources for what Bruno thinks of as a false infinity, he mentions more than once Heraclitus of Pontus both as an early exponent of heliocentricity and as the propounder of a unique universe with outside it an infinity composed of fire. For the astronomical fragments of Heraclitus, see Gottschalk 1980: chap. 4.

infinity as the expression of God's plenitude. If the infinite universe is the direct imprint of the infinite mind of God, then it cannot admit of interruptions between a material world of things and an immaterial world of pure, divine light. God's plenitude must be present and effective in all parts of the infinite whole. In a sense, then, the compromise solution of the post-Copernican sixteenth-century Hermeticists, who tried to fill up an infinite space with a neo-Platonic semimaterial and semispiritual element such as light, seems to Bruno as logically false as the traditional closed universe it attempts to replace. Not even Plato, Bruno notes, dared to suggest that the finite can be continuous with the infinite: incorporeal being cannot exist side by side with corporeal being because they have no common genus. Those who situate "another nature" beyond the visible stars are unable to produce demonstrable proofs and must be dismissed as wishful thinkers and dreamers.[26]

A universal continuum, in Bruno's opinion, must be founded on a homogeneous concept of substance. By the time Bruno wrote these pages of the *De immenso* he had already defined his universal substance in atomic terms in the first work of the Frankfurt trilogy, the *De triplici minimo*. His infinitism can be fully understood only in the light of his reproposal of ancient atomism. Here, however, we are interested in the connection between Hermeticism and universal infinity in the context of the post-Copernican cosmological discussion. What has emerged is not only that Bruno's infinitism was not founded on the Hermetic texts but that those texts were being used by many cosmologists of the period to authorize a compromise solution to the question of a post-Copernican universal infinity of which Bruno himself was highly critical.

The cosmology of Trismegistus remained safely anchored to the ancient concept of a unique universe bound by the circle of the fixed stars. Outside, if there was anything, was a spiritual entity leading the soul to God, the macrocosm, figured as an infinite circle with an unbounded circumference. In the *Asclepius*, Trismegistus urged his young pupil to turn away from the vicissitudes of the sensible world and, aided by the divine gift of reason, attempt to penetrate the higher truths of the spiritual garment that envelopes the sensible universe of man.[27] This Trismegistan dichotomy is echoed in a seventeenth-century text quoted by Furley: Walter Charleton's *Physiologia Epicuro-Gassendo-Charletoniana: or a Fabrick of Science Natural upon the Hypothesis of Atoms*. Although professing atomism, Charleton refuses to consider atoms as the substrata of

26. *De immenso*, bk.VIII, chap. 5. See Bruno 1879–91, I, ii:300.
27. See Copenhaver, ed., 1992:89.

an infinite universe. Rather, he calls piously on Trismegistus to justify a scientific inquiry sanely anchored, in Baconian terms, to the sublunar world:

> If any Curiosity be so immoderate, as to transgress the Limits of this All, break out of Trismegistus Circle, and adventure into the Imaginary Abyss of Nothing, vulgarly called the Extramundane Inanity; in the Infinity (or rather Indefinity) of which many long-winged Wits have, like seel'd Doves, flown to absolute and total loss: the most promising Remedy we can prescribe for the reclaiming of such Wildness, is to advertise, that a serious Diversion of thought to the speculation of any the most obvious and sublunary natures will prove more advantageous to the acquisition of Science, than the most acute metaphysical Discourse.[28]

It seems possible that the reference to "seel'd Doves flown to absolute and total loss" is a colorful but discreet way of recalling Bruno himself, particularly in view of Rita Sturlese's thesis that during his Copernican lectures at Oxford, rudely interrupted by the scandalized Oxford dons, Bruno was reading his Latin *Sigillus sigillorum*.[29] If so, Charleton's text could be considered an early misreading of Bruno; for Bruno does take from the Hermetic texts the idea of the world, even in its most humble material manifestations, as imbued with God-given life and intelligence. This divinization of matter, however, is an idea that Bruno applied rigorously throughout an infinite universe that has no bounds between a "lower" and an "upper" sphere of being: an idea the Hermetic texts refuse. It is therefore incorrect to speak of Bruno as experiencing in an infinite Hermeticized universe "an expanded form of gnosis." There can, in his universe, be no gnosis because no "higher" or "other" world is available to the mind of man. The mind, of course, contains varying degrees of knowledge of the infinite universe, but that is another question. Bruno made his rejection of the idea of a gnosis in his infinite universe explicit in his critical comment on Palingenius in the last book of the *De immenso*: Palingenius, it must be remembered, is praised as the thinker who had got closest to a correct idea of an infinite universe, but not, in Bruno's opinion, close enough. Commenting on Palingenius's conviction that the world derives from an evil principle, whereas God's will, expressed in the divine light that diffuses the universe beyond the band of fixed stars is the essence of justice and good, Bruno writes: "Although it is inappropriate to

28. Charleton's text, published in 1654, is quoted in Furley 1981:576.
29. See Sturlese 1994.

do so, if you wish to affirm that nature is the place of evil, you are obliged to admit that there exist two kinds of principles, and in that way you will gain the approval of the sect of the gnostics."[30] For this reason, Bruno finally put Palingenius aside and, in the final pages of the *De immenso*, celebrated his own idea of an entirely homogeneous, infinite universe, filled throughout with one constantly moving and modifying substance.

Recent criticism, as well as emphasizing the differences between Bruno's infinitism and that of the post-Copernican Hermetic cosmologies such as those of Palingenius, Patrizi, and Digges, has also been much concerned with Bruno's definition of the infinite world through a rigorous logical confrontation with Aristotle and the Christian theologians of the Middle Ages: a subject which had already been excellently treated by Hélène Vedrine in 1967. In a later major study of Bruno's "eclectic scholasticism," carried out in the wake of Charles Schmitt's reevaluation of renaissance Aristotelianism, Paul Richard Blum has underlined the importance for Bruno's intellectual formation of his teachers at the monastery of San Domenico Maggiore in Naples, and in particular of Matthias Aquarius.[31] Referring to Aquarius's annotated copy of the comment by Johannes Capreolus to the *Sentenze* of Pietro Lombardo, Blum has demonstrated that a strong interest in the question of the possible infinity of the universe was already present in the monastic Naples of Bruno's early years.[32] After considering, in the light of late-scholastic discussions, the problem of God's *potentia* with relation to his creation of the world, Aquarius rejects the infinite universe, although the terms of that rejection become increasingly ambiguous as the discussion wears on.

When Bruno reached Paris in 1581, he would already have been well versed in scholastic discussion of the infinite and may have paid particular attention there to the treatise on the subject published at the beginning of the century by the Scots logician John Major, active at the Sorbonne under the name of Jean Mair (1469–1550). In his complete treatise of logic published in 1506, Mair had inserted between a section of *Argumenta sophistica* and a *Dialogus inter duos logicos et magistrum* at the end of the volume, a series of *Propositum de infinito* divided into three questions: Is there an infinite in act? Is it contradictory that God produce an infinite creation? On the movement of an infinite body.[33] There are

30. *De immenso*, bk. VIII, chap. 6. See Bruno 1879–91, I, ii:302.

31. See Blum 1994.

32. For a more general account of the studies that Bruno would have been required to undertake during his monastic period in Naples, see Miele 1992.

33. For a modern French translation, see Mair 1938.

aspects of Jean Mair's treatise which suggest that Bruno knew it well: for example, that *Dialogos inter duos logicos et magistrum*, which takes the form of a neo-Aristotelian debate on formal scholastic propositions similar to those put forward by Nundinius in the central dialogue of Bruno's *Ash Wednesday Supper*. Theophilus's suggestion in the *Supper* that Nundinius himself had "gone into Scotland" may not refer to an actual journey but rather to Nundinius's association with the figure of Jean Mair, who also rejects the arguments in favor of an infinite creation on the part of God.[34]

Mair, however, excepts from this negative judgment an argument that he develops at length. That is the evidence of the spiral line, which he distinguishes from a circular line insofar as it fails to finish where it commenced but rather stretches in a cylindrical form into infinity. The spiral line, as Mair probably knew, was associated not only with spatial infinity but with temporal infinity as well. In a discussion in the pseudo-Aristotelian *Problemata*, the Pythagorean Alcmeon is described as saying that people die because they are unable to join beginning to end, a remark that seems to associate the spiral line with immortality through metempsychosis.[35] The spiral line thus authorized a series of highly unorthodox doctrines from a Christian point of view; and Mair has good reason to associate it with the serpent in the tree of knowledge.[36] When he comes to his final repudiation of the various arguments in favor of an infinite universe, Mair nevertheless admits that the spiral line holds meanings which are too potent for him to contradict, although he enigmatically refuses to explicate those meanings. Bruno himself adopts a very similar attitude when, in a much commented page of the *Heroici furori* the Neapolitan poet Tansillo (a historical figure belonging to a previous generation) also uses the spiral line to illustrate the movement of thought which, metaphysically, leads the mind through the grades of perfection until that infinite center is reached which is neither form nor formed. Tansillo's pupil, Cicada, asks how it is possible, moving in circu-

34. "Però quanto a certa instanzia che produsse Nundinio de gli monti di Scozia, dove forse lui è stato, mostra che lui non può capire quello, che se intende per gli altissimi monti." See Bruno 1958:115.

35. See Sorabji 1988:chap. 10. Sorabji claims that the Pythagoreans and possibly also the Stoics thought about time as circular rather than linear. He further points out that the Great Platonic Year, which corresponded to the period at the end of which all the stars would have returned to their original positions (that is, the period of precession), offered an image of the reduplication of worlds in circular time.

36. ". . . la ligne circulaire si distingue de la ligne spirale en ce qu'elle finit au point où elle a commencé, tandis que la spirale finit au point opposé, comme le serpent dans l'arbre d'Adam": Mair 1938:15.

lar motion, to arrive at the center. Tansillo replies enigmatically that such motion exists, although, when pressed by Cicada, he refuses (as does Jean Mair) to explain his meaning further: "I can only tell you, and leave you to consider it."[37] Here scholasticism has been dissolved in neo-Platonic mysticism following an idea favorable to universal infinity; although it is not yet explicated in Bruno's more realistic cosmological terms.

Bruno's Spanish translator, Miguel A. Granada, has published an important study of Bruno's refusal of the traditional distinction between God's *potentia absoluta* and his *potentia ordinata*.[38] Granada demonstrates how the refusal of this distinction was at the basis of Bruno's extremely personal and original concept of physical infinity. The classical studies of Bruno's infinitism in Arthur O. Lovejoy's *Great Chain of Being* and Alexandre Koyré's *From the Closed World to the Infinite Universe* had emphasized the dependence of Bruno's idea of physical infinity on the logical argument known as the Principle of Plenitude. This argument states that an omnipotent God can only create an infinite world, infinitely full of being, because a finite world would represent a limit to divine power in the act of creation. Lovejoy and Koyré underlined how the Christian theology of the Middle Ages had opposed to this principle a series of restrictions designed to save the idea of divine free will in the act of creation: God freely decides to restrict his divine powers in the creation of a finite world in a definite moment in time. Granada, following an indication put forward by Alfonso Ingegno in his studies of Bruno's religious thought, points out that these theological restrictions were based on the conceptual distinction between *potentia absoluta* in God and *potentia ordinata*.[39] These were not defined as two different divine powers, but rather as two different ways of conceiving of divine omnipotence. God's creation, by virtue of his *potentia ordinata* of a passive, finite material world through which he acts as the giver of forms, is the way he has chosen to express his *potentia absoluta*; and the distance between the two represents the possibility of divine intervention in the material world through the power of miracles.

The Hermetic forms of an infinite universe that preceded or were contemporary with Bruno denied this distinction, claiming that God's *potentia absoluta* required him to create an infinite universe. Going back to ancient Platonic and Aristotelian hierarchic concepts of being, however,

37. See Bruno 1958:1012–3. This passage was the subject of some of the most famous pages of nineteenth-century Bruno criticism. See Bertrand Spaventa, "La dottrina della conoscenza di Giordano Bruno," *Saggi di critica*, 1867, 252–55.

38. See Granada 1994.

39. See Ingegno 1984 and 1987.

the Hermeticists filled up the higher regions of their infinite space with some divine substance such as light or intelligence, maintaining the idea of a unique material universe of finite dimensions. Bruno, as Granada demonstrates, expresses a far more radical refusal of the scholastic distinction between *potentia absoluta* and *potentia ordinata*. For Bruno, an all-powerful God can, as a logical necessity, express himself only in terms of a *potentia absoluta*, and therefore his entire creation will partake of the infinite, in this world and in all other worlds that exist within the infinite whole. There will be an end to the hierarchy of being. There will also be no more miracles because everything already corresponds to the *potentia absoluta* of God, and what appears inexplicable and marvelous will be due to the limited powers of the mind, anchored as it is to a specific place in space and time. Bruno insists on this motive with particular emphasis in chapter 1 of book I of the *De immenso*: "It is in the nature of humankind, and part of everyone's desire, to be able to embrace totality: everyone desires to be that which he is only at times; to see everywhere what he only sees in particular places; to consider in its universality what only appears to him in its singularity; to make full use of that which he can only use in part."[40] This cluster of ideas raises the problem, on which Bruno was not always consistent, of the existence of universal scientific laws: for example, the movement of all earths in the universe about their central suns.

Granada points out that Bruno's radical refusal of the scholastic distinction between the two forms of divine *potentia* leaves him with tricky problems to solve, the major one being how to explain the evidently finite dimensions of all that we experience in the phenomenological world. Granada's reply to this question concentrates, in my opinion, on one aspect only of Bruno's explanation, thus depriving his thought of much of its originality and complexity. The aspect which Granada correctly underlines is the logical distinction between divine *complicatio* and divine *explicatio* on which Bruno insists in his discussion of divine omnipotence. Bruno accuses the scholastic and Christian tradition of propounding an anthropomorphic concept of God, insisting on attributing to an omnipotent divinity such concepts as will, choice, and liberty, which are relevant only to finite minds. He claims instead that the absolute quality of God's infinity and power cancels such concepts. In God, choice and necessity, liberty and act, are contemporaneous and identical. It follows that in the world created by God's *complicatio*, all those elements which, in his *explicatio*, appear finite to our finite minds contain within them also

40. See Bruno 1879–91, I, i:202.

the infinity of God: "I say that God is *completely infinite*, because he excludes from himself every limit, and every one of his attributes is one and infinite; and I say that God is *totally infinite*, because his totality is in all the world, and in every part of it infinitely and totally."[41] Granada recalls Bruno's dependence here on the thought of Cusanus. Bruno, however, modifies Cusanus's doctrine by insisting that the material expression of the totally infinite is not assumed by the mediating figure of Christ, but rather by the infinite universe itself, which becomes the Word, a second God, the only mediator between the human mind and God.[42]

If Bruno's defense of his infinite world were to rest only on this logical concept of the *coincidentia oppositorum* in the mind of God, it could justly be accused of resting on a purely metaphysical basis. Much more physical consistency derives to it from the atomistic argument put forward by Bruno from the time of his Italian dialogues but reelaborated in the first work of the Frankfurt trilogy, the *De triplici minimo*, as the foundation of his concept of universal infinity as it will be defined in the *De immenso*. For the central characteristic of Bruno's atomism is the idea that every atom contains all the power and virtue of divine infinity. The atom is thus the ultimate receptacle throughout the infinite universe of the *potentia absoluta* of God: a concept that allows Bruno to identify the creative power of God with the maximum unity of the infinite universe and to conceive of both of them as infinitely compressed within each minimum unit or atom in a page which, as I have shown elsewhere, was being eagerly read in England by the scientists surrounding the ninth earl of Northumberland.[43]

> And yet we must affirm that finite matter, however extended, is not composed of infinite parts, but if we continue adding to it, like adding innumerable numbers to a finite number, then it really is possible to proceed to infinity. On the contrary, by subtracting and dividing the parts of a finite magnitude, we necessarily arrive sooner or later at the minimum quantity, just as by subtracting a number from a finite number we must arrive at the monad.[44]

And it was surely because Bruno had developed this "coincidence" between the infinitely small and the infinitely great that Harriot, in a letter

41. See *De l'infinito, universo e mondi*, in Bruno 1958:382.

42. For Bruno and Cusanus see Vedrine 1970, the relevant pages of Bernart 1986, and Ingegno 1987:113–33.

43. See Gatti 1989:60.

44. See *De triplici minimo*, bk. I, chap. 6, in Bruno 1879–91, I, iii:152–53; my translation.

of 1608 to a disapproving Kepler, could write his much-quoted passage: "I have led you to the doors of nature's mansion, where her secrets are hidden. If you cannot enter on account of their narrowness, abstract yourself mathematically, and contract yourself into an atom, and you will enter easily. And after you have come out, you will tell me what wonders you have seen."[45]

Bruno's atomism solved for him the problem of explaining the apparently finite aspects of an infinite world. For if every aspect of being consists of a meaningful agglomeration of atoms, and if each minimum atom contains all the maximum power of God, then all matter is imbued with the absolute power of the divinity and finite only according to the subjective limitations in space and time of the perceiving qualities of the human mind.

> It is a long time since the death of a single time for all things, subjectively located in the highest heaven. . . . So, you will draw your conclusions with respect to the moments of time and the phases of movement, not starting from the point of view of the subject, given that the boundaries of one genus do not define the boundaries of another genus, and the time of one genus cannot be considered as unique with respect to the time of another.[46]

There were good reasons why post-Kantian philosophers such as Schelling, Coleridge, and Hegel were reading the works of Bruno.[47]

45. The correspondence between Kepler and Harriot was conducted in Latin. The entire exchange has been published in Kepler 1937–75, vols. 15 and 16. The English translation in my text is from Jacquot 1952. Kepler's correspondence with Harriot is also discussed in Shirley 1983:385–88. The significance of Bruno's idea of the atom as not only a minimum quantity but at the same time a maximum source of energy was pointed out by Paul-Henri Michel in two important studies of Bruno's atomism. See Michel 1957 and 1964.

46. See *De triplici minimo*, bk. I, chap. VIII, in Bruno 1879–91, I, iii:168; my translation. For an interesting comment on Bruno's treatment of time in the *Camoeracensis acrotismus*, see Capek 1987; and for a more thorough and profound if less scientific treatment of the same theme, Badaloni 1997.

47. See Beierwalters 1973/1983, Blumenburg 1975/1987:353–85, and Ricci 1991 for Schelling and Hegel. For Coleridge and Bruno, see Gatti 1996.

7 The Infinite Worlds

Developing a scholastic distinction, Miguel Granada has extended his important work on Bruno's infinite *extensiva* to his infinite *intensiva*, located in the infinite number of celestial worlds. Here and in the following chapter I suggest that Bruno's infinite *intensiva* was located ultimately in the infinite number of indivisible atoms, although he never abandoned the special status, accorded in the Italian dialogue *On the Infinite Universe and Worlds*, to the innumerable worlds as privileged agglomerations of atoms, which he thought of as infinite in number although finite in kind. The fact that finite body and finite dimension are the ways in which the human mind understands the physical universe must not, in Bruno's opinion, lead to the mistaken idea that finite dimensions are not real. The finite aspect of bodies is very real; for it is the way in which the divine intelligence intends them to be understood. Equally, the divine intelligence intends the universe as a whole as essentially infinite (both *extensiva* and *intensiva*), and that means that it really is infinite, and must be so. The infinity of the whole, Bruno writes in the second dialogue of *De l'infinito*, is even more necessary to a proper understanding of the universe than the finite nature of its parts.[1]

A necessary consequence of Bruno's thought about the principle of plenitude is that the infinite universe cannot be empty, either inside or

1. "... come realmente si trovano corpi dimensionati finiti, cossì l'intelletto primo intende corpo e dimensione. Se lo intende, non meno lo intende infinito; se lo intende infinito ed il corpo è inteso infinito, necessariamente tal specie intelligibile è; e per essere produtta da tale intelletto, quale è il divino, è realissima": Bruno 1958:395.

outside our own world. Although Bruno will sometimes describe his infinite space as nothing more than an infinite container subject to a tridimensional Euclidean geometry, it is in fact never, and in no part, to be seen as an empty container.[2] For Bruno fills his universe with an infinite material substance, which he calls by the traditional term "aer" or "ether." This ether must be continuous with the infinity of the universe itself; for, as we have seen, Bruno objects to any discontinuity within the infinite whole. How, then, is he to account for atoms? In the second book of the *De immenso*, Bruno approaches the problem in these terms: "The single atoms and species are limited, but all the atoms (according to the number of the indefinite average species and the prime definite species, which are the elements) are infinite and infinite is matter itself out of which all things are formed, and infinite is space in which all things are contained and limited, endowed with a finite activity and passivity."[3] Bruno places the infinite number of single atoms in an infinite ether which is identifiable with infinite space, and as such should be considered as logically prior to bodies in space. Within the all-pervading ether, an infinite number of touching minimums, or atoms, all containing a total divine potency, and not a part of that potency, emerge into finite forms like the formation of water drops from the immensity of the ocean. A first formation is seen as "prime" and gives rise to the traditional "elements." From these a second process of formation gives rise to discrete bodies. In Bruno's universe, however, no composite body is ever completely discrete. The occasional agglomerations of atoms, through constantly changing combinations of the elements, are always unstable, forming and reforming, condensing and dissolving.[4]

The active forming principle in this universe of constant vicissitude is located within the infinity of the infinite substance, its forming activity visualized by Bruno as a kind of universal yeast. This forming principle is identified by Bruno with soul. All soul in Bruno's universe is divine and therefore intelligent. There are no vegetative or even sensitive souls cor-

2. "Space is a certain continuous physical quantity, in three dimensions, which contains everything indifferently . . . it stretches beyond everything and includes everything. . . . Space is not movable but gathers within itself the things which move": *De immenso*, bk. I, chap. 8, in Bruno 1879–91, I, i:231–32.

3. *De immenso*, bk. II, chap. 9, in Bruno 1879–91, I, i:291; my translation.

4. See in particular bk. V, chap. 1, of *De immenso*. Here Bruno doubts the ability of the rational finite mind to follow the intricate process of constant change and vicissitude which characterizes the infinite whole. He dwells in particular on certain inconstant forms such as the goat "elongated" by the kids that follow in her wake, or, once again, the spiral line that finds its image in the serpent with its shining, slippery back. For this aspect of Bruno's thought, see in particular chaps. 5 and 6 in Badaloni 1988:69–95.

responding to the traditional Aristotelian hierarchy. The soul of a stone, with respect to that of an apparently higher body such as an animal or man, is not different in kind. It simply has a less fully articulated form than others, through the particular disposition of its atoms, of the substrata of infinite substance. Even that may be an erroneous judgment depending on our incapacity to understand the particular life form of stones. In book V, chapter 1, of the *De immenso*, Bruno insists on the negative consequences of attempting to determine the nature and measure of all things on the basis of human criteria. And he mentions in particular "certain stones" which are imbued with a particular force of attraction that defies the human senses: a clear reference to the magnetic properties of iron.[5]

It follows from these premises that Bruno will fill up his infinite universe with an infinite number of worlds. These are a logical consequence of the infinite number of atoms which, impregnated with what Bruno sometimes calls soul, sometimes a "vital fluid," constantly coalesce and dissolve into an infinite number of forms. His debt here to the ancient atomists is explicit and clear. In the final *De immenso*, he deliberately writes a Latin scientific poem in imitation of the Lucretian *De rerum natura*. Bruno's reading of Epicurean and Democretean atomism is cautious and critical, however. In the cosmological context, his criticism is especially evident in his treatment of the theme of the disposition in space of the infinite number of worlds. Referring to what he calls a "stupid" opinion attributed to Epicurus by both Diogenes Laertius and Lucretius, Bruno denies that the immense universe contains only one sun.[6] For Bruno, the logical consequences of the atomistic explanation of matter include not only an infinite, homogeneous substance as the basis of universal being, but also the formation of an infinite number of celestial bodies of a substantially similar nature throughout the infinite whole. Bruno calls these infinite worlds angels or gods insofar as they mediate between the finite human mind and the universal, infinite whole.

All the celestial bodies in space, including earth, are, in Bruno's opinion, closely related. With respect to the casual agglomerations of atoms, and the consequent casual disposition of bodies in space of the ancient atomists, this concept of an intimate universal order had been one of the keystones of Aristotelian cosmology and physics. Bruno recognizes its force and mediates through Aristotle's Arabian disciple Averroës his idea of matter as a permanent principle that, far from assuming its forms in a

5. See Bruno 1879–91, I, ii:114–5.
6. See *De immenso*, bk. II, chap. 9, in Bruno 1879–91, I, i:290.

passive, casual sense, constantly produces them from an internal intelligence dominated by an idea of order.[7] Yet Bruno knows that he must find a new conceptual foundation for the order of the cosmos, for he has already deprived his own universe of both a prime mover and of the idea that all motion is centered on earth. One possibility would have been to fall back onto a Christianized theological concept of universal order; and for this there was an authoritative model in the work of Nicholas of Cusa, whom Bruno read and admired.

Bruno's interest in the philosophy of Cusanus was probably owing to the clarity with which the fifteenth-century cardinal started his inquiry from the same problem that is at the source of so much of Bruno's cosmology. In his best-known work of 1440, *On Learned Ignorance*, Cusanus underlines the incommensurability between infinite and finite wisdom. According to Cusanus, divine and cosmic infinities remain out of line with man's finite conceptions. This intuition led Cusanus, in the second book of *On Learned Ignorance*, to propose a cosmic relativism that rejected the Aristotelian-Ptolemaic cosmology. For Cusanus denied the one universe centered on earth, claiming that the shadow and dark color of the earth could not justify its being placed at the bottom of universal being as the place toward which all things gravitated. Rather, Cusanus claimed that the sun was made of the same elements as earth and that throughout an infinite universe, without circumference or center, all celestial bodies are suns representing to the same extent the explication of God's creative power. Yet even when he had filled an infinite universe with divine light and had further proposed (with neo-Platonic echoes) an infinitising geometry based on the identity between the minimum arc and the minimum cord, or the infinite line and the maximum circle, Cusanus was unable to bridge the final gap between a human idea of infinity and the divinity of an infinite infinite beyond our understanding. It was on the basis of this intuition of a fundamental discontinuity and incommensurability that Cusanus justified the necessary incarnation of Christ.[8]

Much has been written on Bruno's open admiration of the "divine Cusanus" and his even more frequent, often unacknowledged, quotations from his works.[9] Bruno's cosmology has even been explained as a fusion of Copernicus's heliocentricity and the infinitism of Cusanus. We must

7. For Bruno's reading of Averroës, see Sturlese 1992a.

8. For the thought of Cusanus and its influence on Bruno, see Vedrine 1970.

9. See Tocco 1889:60, Cassirer 1927/1963, Ingegno 1987:113–48, and Bruno 1991a, where Sturlese underlines Bruno's use as a source not only of *On Learned Ignorance* but also of *De beryllo*.

not, however, overrate the influence of Cusanus. Although Bruno incorporated some important elements from Cusanus into his cosmology, his final estimate of Cusanus's work was as critical as his reading of, for example, the ancient atomists. The most articulate expression of Bruno's reading of Cusanus can be found in the third dialogue of *De l'infinito, universo e mondi,* where the subject of discussion is the matter out of which the sun is made with respect to the earth. Quoting from *On Learned Ignorance,* Bruno refers approvingly to Cusanus's doctrine of the essential similarity between the sun and the earth, composed, although in differing proportions, of the same elements. Bruno recognizes that Cusanus has already destroyed the Aristotelian cosmology, with its rigid distinction between the higher and lower regions of the universe, where he calls him "one of the most remarkable geniuses that have ever breathed under this air." He goes on to quote a further conviction of Cusanus that the earth and the moon, as well as the sun, should be considered luminous. Bruno says this is a doctrine which Cusanus could have done without, for he is filling his infinite universe, which is no longer divided into higher and lower spheres, with an infinite substance which itself is divided into two radically different elements: a subtle, ethereal, fiery matter which he considers an appropriate emanation of the divinity, and the more clearly material parts of the universe, such as the opaque face of the moon, which are fundamentally discontinuous with the divine light. Bruno comments that Cusanus, in his infinite universe, was like a swimmer struggling through tempestuous waves, sometimes high and sometimes low, with the added inconvenience of sudden interruptions between the crests of the waves and their troughs. Bruno instead proposes a serene and tranquil swim through a homogeneous infinity composed of a continuous substance within which fiery hot bodies, and cold opaque ones, are to be considered as different formations, though both of them ultimately derive from the unique substance that stretches out into infinity.[10]

Through this discussion of Cusanus, Bruno brings into prominence (again with shades of Pythagoras) a fundamental pair of contraries active throughout the infinite universal whole. These are fiery, hot substances and cold opaque substances. For Bruno, the genius of Cusanus lay in his perception that this dichotomy was not to be considered, in Aristotelian terms, as dividing the universe horizontally into two: a fiery celestial quintessence "up there" and a cold earthly solidity "down here." The fundamental dichotomy, for Cusanus, was discernible throughout

10. See Bruno 1958:440–42.

the infinite whole. This intuition deprived the universe of a center, making any center purely relative as a point of observation. His dichotomy, however, in Bruno's opinion, kept Cusanus at a distance from his transcendent God; for he was unable to accept that his universal dichotomies were founded on a principle of immanent universal unity from which they were constantly born and to which they constantly returned. Cusanus was therefore obliged to bridge his logically unbridgeable distance between unity and duality with reference to the incarnation of Christ: for only a miracle could resolve the unresolved play of contraries and contradictions in the phenomenological world. Bruno could solve his cosmological problem without reference to the Christian incarnation or to miracles, for he thought of his infinite, eternal substance as a direct explication of the absolute power and unifying principle of God.

By associating the two fundamental Cusanian contraries, light and shadow, with another pair of contraries in the form of heat and cold, Bruno was turning away from Cusanus toward the works of Telesius. In chapter 16 of the first book of his *De rerum natura*, Telesius had complained bitterly of the difficulties he had encountered in the publication of his work, which he must have found all the more galling as he had stretched his orthodoxy to its limits in an attempt to avoid the censures of the Catholic Church. Nevertheless, in the preface to his work, Telesius had declared himself willing to detract, if required, whatever the authorities might wish.[11] Furthermore, he never suggested that his doctrine of matter should be applied to any universe except the accepted one of Aristotelian-Ptolemaic fame. Nevertheless, the title of his work was provocative; and it must have been clear that his fundamental thermodynamic dichotomy could be extended throughout the material universe and not only applied to that part of elemental matter which lay beneath the lunar sphere.

Bruno took over Telesius's two "natural agents," heat and cold, in more strictly Lucretian terms than Telesius had dared to propose, in spite of the Lucretian title of his major work. Bruno applied them throughout to a universe of infinite dimensions composed of a homogeneous, material substance that founded the play of contraries in a unifying sense of the infinite whole. This use of Telesius's two basic natural agents, heat and

11. The two books of Telesius's *De natura iuxta propria principia*, published at Rome in 1565, were reissued in Naples with the title *De rerum natura* in 1570. The final text of Telesius's work, in nine books, was not published until 1586, in Naples. Although Bruno could not have known the complete work when he published his Italian dialogues in London between 1584 and 1585, Aquilecchia believes he had read it by the time he published his Frankfurt trilogy in 1591. See Aquilecchia 1993a:298.

cold, led Bruno to his most original contribution to modern cosmology: not only the idea that a universe of stars lay beyond our own sphere of outer stars but that these further stars must themselves be ordered into systems similar to our own system of planets revolving around a central sun. Bruno divided all celestial bodies into hot or luminous bodies, or suns, and cold opaque bodies that revolved around them to obtain their life-giving heat and light. He thought that all stars were suns, with bodies revolving around them so small that the eye could not detect them. Although this idea can hardly be said to correspond to the modern galaxy, it was certainly getting close. Bruno already thought not only that other worlds or star systems existed outside our own, stretching toward infinity, but that those worlds must be modeled on principles similar to the ones that defined the order of our solar system.

The thermodynamic principle, mediated through Telesius, which united all the infinite worlds into coherent order and forms was of great importance to Bruno because it supplied him with a biological explanation of the movements of an infinite number of worlds throughout an infinite space. To discuss Bruno's cosmology without reference to the thermodynamic principle mediated through Telesius leads to what appears to me the serious error, so often repeated in Bruno criticism, that his infinitism was not a properly cosmological concept but only the place in which to experience a kind of religious *enthusiasmus*, or perhaps a more rational contemplative neo-Platonic ascent of the soul. This interpretation has a long history, going back at least to the puzzled reading of Bruno's post-Copernican cosmology by nineteenth-century scientific commentators such as Schiaparelli, referred to by Felice Tocco, who could not understand how a properly astronomical argument could be expressed in the exuberant mythical and metaphorical language so often used by Bruno.[12] It received strong endorsement from the Hermetical school of Bruno scholars, and in particular from Frances Yates, whose 1964 book significantly makes no mention of Telesius's thermodynamic theory of matter.[13] Alfonso Ingegno and Miguel Granada have also been

12. The distinguished astronomer Giovanni Schiaparelli was consulted by Felice Tocco to illuminate Bruno's account of the movements of the earth and replied with a letter that Tocco published in his work comparing Bruno's Latin texts with the Italian. See Tocco 1889:313–35. Ingegno 1978:64 has pointed out that Schiaparelli's letter, rather than clarifying Bruno's reading of Copernicus, tended rather to add to the confusion by suggesting that, although Bruno certainly intended to supply an interpretation of Copernicus's astronomy, what he came up with was devoid of any serious astronomical basis.

13. Rita Sturlese's reading of Bruno's philosophy as a modified, immanentistic version of Ficino's neo-Platonism writes off with equal haste both Bruno's Copernicanism and his reference to Telesius. See Sturlese 1994.

considering the important point of Bruno's refusal of the traditional distinction between the *potentia absoluta* and the *potentia ordinata* of God, that is, his absolute power to create eternally in an infinite number of ways and his power to create in a moment of time the finite universal order in which we live.[14] Once the infinite universe is considered by Bruno as the space in which God's *potentia ordinata* and his *potentia absoluta* coincide, the infinite material substance itself becomes imbued with the total power of the divine intelligence. Bruno uses both neo-Platonic and Hermetic sources and vocabulary to express his sense of the divinity of his infinite universe; and any full account of his philosophy would have to consider his significant contribution to the lively renaissance discussion of metaphysical questions such as the nature of death, the problem of time, and the immortality of the soul. It is a mistake, however, to separate the consideration of such aspects of his thought from his natural philosophy. Precisely because Bruno sees the infinite material substance as rationally ordered by the divine intelligence, its infinite extension must necessarily assume a coherent physical ontology which he expresses as an infinite number of worlds.

Bruno's infinite universe is conceived of as an intelligible cosmology in the modern scientific sense of the term. The laws controlling that intelligibility would be defined in the following century in increasingly refined and complex terms. Nevertheless, Bruno already proposes an unlimited cosmos in which motion assumes a basic ordered form, that of the movement of colder bodies around suns, mediated through the heliocentric astronomy of Copernicus. It is a cosmos in which that movement finds a theoretical physical foundation in the concept of an infinite and eternal opposition of contraries, mediated from Pythagoras through Cusanus. It is furthermore a cosmos in which one basic opposition, that of heat and cold mediated through Telesius, provides the explanation of planetary movement throughout the infinite whole. That Bruno's sense of the immensity of his infinite universe, as well as the complexity of the infinite vicissitudes to which it is necessarily subjected as the receptacle of God's absolute power, led him to develop a problematical attitude toward modern mathematics, and to the epistemological problem generally, is clear, and will be the subject of the following chapters. Nevertheless, the importance of his attempt to define the intelligibility of his infinite universe with its infinite number of worlds in terms of a new physics must not be ignored.

14. See in particular Granada 1994.

Bruno's attempt to apply the thermodynamic philosophy of Telesius to an infinite post-Copernican cosmology, foreign to Telesius himself, soon showed up its deficiencies. Telesius, as we have seen, failed to supply Bruno with the explanation of some planetary movements, such as that of the moon around the earth, which remained a puzzle in his system until the end. Although Bruno finally, in the *De immenso*, accepted that Copernicus had placed the moon on an epicycle revolving around earth, there was no real reason in Bruno's cosmology why that should be so. The magnetic philosophy of Gilbert, which would later be preferred by Galileo, offered a better explanation of the movements of bodies in the ether because it made them dependent on relative magnetic forces related to their mass. It is nevertheless of considerable importance that Gilbert himself mediated much of his cosmological reasoning through Bruno.

Johannes Kepler was far less enthusiastic than Gilbert about Bruno's cosmology. Kepler admits that the idea of a universal infinity populated by infinite worlds, which entirely destroys any idea of a privileged place for our own world within the universal whole, left him distressed and without any clear idea of his mental bearings: an attitude clearly shared by Tycho Brahe, who dismissed Bruno curtly as a Nolan *nullus*.[15] Kepler, however, discussed Bruno's cosmological speculations very seriously, even if he disagreed publicly with the most forward-looking and original aspect of his cosmology: his concept of universal infinity and the idea of an infinite number of solar systems within unbounded space. In chapter 21 of his *De stella serpentari* published in 1606, Kepler argued that our universe, which he thought of as post-Copernican and centered in the sun, fills the whole of space. He added more acutely that the fixed stars we see are related to the central sun and ought not to be considered suns of distant solar systems similar to ours. In his later *Dissertatio cum nuncio sidereo* of 1610, a public discussion with Galileo after the latter's discovery of the moons of Jupiter, Kepler refers frequently to Bruno's infinitism: a subject, he informs Galileo, much in discussion in the imperial court in Prague as a result of the enthusiastic adherence of the influential nobleman Giovan Matteo Wackher to a Brunian multisolar and infinite cosmology. Kepler had feared that the discovery by Galileo of previously unknown bodies revolving within the universe would confirm Bruno's theory of multiple solar systems; but he is comforted by the news that the discoveries are limited to the moons of Jupiter and therefore remain within our own and, for Kepler, unique solar system. Bruno's theory of

15. Kepler's extended reference to Bruno is discussed in Field 1988. See also Ricci 1990:69–76. For Tycho Brahe's disdain for Bruno's cosmology, see Sturlese 1985.

8. Diagram in Johannes Kepler, *Epitome astronomiae* (1618). Reproduced by permission of the Biblioteca Nazionale Centrale, Rome.

multiple solar systems within an infinite universe populated by infinite worlds will appear valid, writes Kepler to Galileo, only if and when future observers equipped with instruments of exceptional precision sight other suns similar to ours.[16]

In his later textbook of post-Copernican astronomy, *Epitome Astronomiae Copernicanae*, published between 1618 and 1621, Kepler conducts another discussion of universal infinity: a cosmological concept which is developed by an alter ego and illustrated in Figure 8. The text relating to this illustration runs:

> If nothing more certain than this were known about the fixed stars, it seems as though their domain were infinite: and this Sun of ours no more than one

16. On the discussion between Kepler and Galileo after the publication of Galileo's *Sidereus nuncius*, see Chevalley 1984.

> of the fixed stars, larger and brighter for us only because it is nearer than the fixed stars are; and so round any of the fixed stars there might be a world such as that round us; Or which comes to the same thing, among the innumerable places in this infinite collection of fixed stars, this our World with its Sun will be only one, in no way different from the other places near other individual fixed stars: As shown at M in the adjoining figure?

Kepler's reply to his alter ego is that it does not follow that because the centers of the stars do not all lie in one spherical surface then their distribution in space must be uniform throughout infinity—a uniformity that is not a part of Bruno's argument for infinity. Bruno does not claim that all solar systems are identical or that they lie at uniform distances one from the other. His argument, which is based on the concept of necessary plenitude, is that the bodies in infinite space must themselves be infinite in number although finite in kind. Furthermore, as, following Telesius, all celestial bodies are either hot or cold, luminous or opaque, throughout the infinite whole the cold and opaque bodies will necessarily circulate around the hot and luminous ones, in order to guarantee the infinite and eternal process of generation and corruption, that is, the infinite process of life. The homogeneity of Bruno's universe should therefore not be confused with the neo-Platonic, geometrical uniformity which Kepler considers a cosmological principle in pages which demonstrate above all a distaste for cosmological speculation about the shape of things beyond the limits of what can actually be observed.

Kepler's critique of Bruno's infinitism was not only directed against Whacker. In a letter of November 5, 1603, the astronomer Edward Bruce had written to him declaring his belief in infinite worlds which move through a space filled with an air similar to ours.[17] These cosmological premises were also accepted in England by the circle of scientists linked to the ninth earl of Northumberland, which included Thomas Harriot, who corresponded with Kepler. Harriot and his friends, however, developed their ideas on cosmological infinity through a critique of Kepler in an episode that is of particular interest because it involves one of the first documented English readings of Galileo's *Starry Messenger*.[18] It is not

17. Quoted in Ricci 1991:68. In the *Dissertatio cum Nuncio Sidereo* (see Kepler 1937–75, 6:289) Kepler associates Bruce with the infinite cosmology of Bruno, defining Bruce as a personal friend of both his own and Galileo's. On the same page, however, William Gilbert considers Kepler to have followed the ancient cosmologer Melissus in proposing a unique but infinite universe. See Ricci 1991:75.

18. For the details of this episode see Jacquot 1952:175–76, Gatti 1989:56, and Ricci 1990:76–79.

necessary here to repeat the details of that episode or to follow up the subsequent seventeenth-century cosmological discussion regarding the idea of universal infinity, which has already been well documented by Saverio Ricci.[19] What must be emphasized, however, is that Bruno's arguments about the infinity of the universe and the infinite number of worlds supplied, at the beginning of the seventeenth century, a point of innovation and discussion, both in England and abroad, which suggests once again that it is unwise to dismiss his contribution to the contemporary scientific debate as marginal or null.

Before leaving Bruno's cosmological speculation, a word needs to be said about his radically eclectic method of procedure. Bruno was no observational astronomer, and his cosmological data derive largely, in humanistic terms, from his reading.[20] If he nevertheless achieves at times startlingly original results, it is because of the width of his frame of reference added to the coherent and independent choices of what to discard and what to use. Bruno's cosmology depends on sources extending back to the ancients such as Pythagoras and Lucretius, through the scholastic debates of the Middle Ages, by way of the fifteenth-century Cusanus to the *novatores* such as Copernicus, Telesius, and Tycho Brahe. The system of natural philosophy which Bruno created from these sources, although often praised for its audacity, is equally often condemned as being shaky and ramshackle. On the contrary, it possessed a remarkable internal coherence and was one of the first modern cosmologies to attempt an explanation of the movements of the worlds through space which could be applied uniformly to an infinite number of worlds in movement throughout an infinite ether.

Bruno's use of the thermodynamic principle as the key to the universal movement of the worlds was only partially successful. Gilbert's magnetic explanation was better, as Galileo would soon point out. But Gilbert clearly had read Bruno. In the context of the post-Copernican search for a physical, explanatory principle regulating the movements of the worlds in space, it is thus possible to trace a line of development from Bruno's universal thermodynamic principle through Gilbert's magnetic philosophy leading to the gravitational laws of Newton.[21] Bruno himself might have been uneasy at the idea that the movement of bodies through space could be expressed by a mathematical law, a question on which Gilbert

19. Ricci 1990:131–238.

20. For the continuing importance of a humanistic methodology even among the most advanced of the new scientists of the period, see Gatti 1993:5–9.

21. On this subject, see also Mendoza 1995.

too was hesitant and uncertain. Nevertheless, it is important to recognize that Bruno was the first to react to the collapse of the Aristotelian cosmology and physics with the realization that it had become necessary to supply not only new measurements and mathematical calculations but also, and perhaps above all, explanations of the movements of an infinite number of worlds which would be valid throughout a universe of unified space.

8 "The Minimum Is the Substance of All Things"

Bruno's participation in the revival of ancient atomism, which gained momentum in the sixteenth century to become one of the dominating motives in the natural philosophy of the seventeenth, has received less critical attention than his infinitism. This may well be because his atomism figures as the cornerstone of a theory of matter that was anomalous in the culture of his time: an anomaly that nevertheless is justified as the coherent foundation of the infinite universe he sought to propose. This essential link between Bruno's infinite cosmology and his development of an atomic theory of matter was acutely noticed by Thomas Kuhn in his few pages on Bruno in *The Copernican Revolution*.[1] Kuhn presented Bruno as the first post-Copernican philosopher to have understood that the full implications of the new astronomy required a parallel revolution in the field of matter theory. Later studies by Paul Henri Michel, Robert Kargon, Carlo Monti, and others have done much to clarify Bruno's participation in the revival of ancient atomism.[2] In this chapter I use their contributions to develop the link proposed by Kuhn between Bruno's particular form of post-Copernican infinitism and his neoatomic theory of matter within the universal whole.

Commentators have noted that the earlier Italian dialogues contain only brief and passing references to an atomic theory. As a result, some of them propose Bruno's atomism as a late development which will find ad-

1. See the section on Bruno in Kuhn 1957.

2. See in particular Michel 1957:249–64, the relevant pages in Kargon 1966, and Bruno 1980:11–27.

equate expression only in the *De triplici minimo*, published as the first work of the Frankfurt trilogy in 1591.[3] Both Michel and Monti contradict this thesis, maintaining that Bruno's atomism must be considered an essential component of his philosophy from the beginning, dating from the composition of the Italian dialogues between 1584 and 1585.[4] They give precise textual references to an atomic theory of matter both in *On the Cause, Principle and One* and in *On the Infinite Universe and Worlds*, but such passages are passing and brief. Tocco was correct, in my opinion, to stress that only with the *De triplici minimo et mensura* does Bruno dedicate an entire work to his atomic theory of matter. Furthermore, he places it at the beginning of the Frankfurt trilogy to indicate that his atomism is now to be considered the foundation and justification of the infinite universe which he will celebrate in the final *De immenso*.[5] The principle of infinite unity has not been denied in favor of atomic multiplicity, although now it can be understood only in relation to the minimum or primary monad in which it finds its dialectical contrary.

All renaissance revivals of ancient atomism went back to the pre-Socratic argument about whether being was one and at rest (Parmenides) or whether being was involved in a continual process of change and becoming (Empedocles, Anaxagoras). Plato's attempt to integrate these two apparently contradictory doctrines was expressed in *The Sophist*, where the true philosopher "like a child begging for 'both' . . . must declare that reality or the sum of things is both at once—all that is unchangeable and all that is in change."[6] Aristotle in book I of the *Physics* spoke sourly of those, like his master, who gave way to both arguments; for he saw that if, on the one hand, they accepted that being means one thing only, on the other they conceded being to "what is not" by accepting the decomposition of composites into atomic magnitudes.[7] Aristotle himself fought his

3. The theory was put forward with great clarity by Felice Tocco in his still essential study comparing Bruno's Latin works with the Italian and has found numerous followers. See Tocco 1889.

4. See in particular Michel 1957:253–54 and Bruno 1980:11, n. 5.

5. The *De immenso* was the final work only of the Frankfurt trilogy. Bruno's final work to be published before his arrest and imprisonment after his return to Italy was a work on the art of memory, the *De imaginum, signorum et idearum compositione*.

6. Plato 1961:994.

7. Aristotle, *Physics*, bk. I, 187a1–10: Some thinkers did, in point of fact, give way to both arguments. To the argument that all things are one if being means one thing, they conceded that what is not is; to that from bisection, they yielded by positing atomic magnitudes. But obviously it is not true that if being means one thing, and nothing can at the same time both be and not be, there will be nothing which is not; for even if what is not cannot *be* without qualification, there is no reason why it should not be something or

long battle against atomism in the name of being as particular things, made up of matter and form, which decompose into a continuous prime substance, an idea that allowed him to contemplate, at least potentially, the infinite divisibility of matter. It is true that Aristotle saw generation in the animal and vegetable kingdoms as depending on corpuscular seeds that carry within themselves the characteristics of the thing which depends on them for generation.[8] To Aristotle, these seeds or "first principles of generation" are "first things" from which a new whole arises. Even the parts of the new births may be seen as developing from their own seeds, which ensure, for example, that the parts of humans or animals resemble the same parts of their parents or forebears. Nevertheless, the seeds are not atoms, as they too can be divided or decompose. Aristotle's doctrine of the infinite divisibility of matter ensured that the final minimum part could never be reached, even if he posited such infinite divisibility as potential rather than actually possible in nature.[9]

In spite of the importance given to seeds in vegetable and animal generation, Aristotle's matter theory rested firmly on the four elements combining in various forms in the sublunar sphere. It further defied the Parmenedian doctrine of the one by positing an entirely different quintessence as the eternal and unchanging matter of the heavens above the sphere of the moon: a subject which, after the publication of Copernicus's *De revolutionibus*, became one of the major cruxes of cosmological discussion in the second part of the sixteenth century. On one side, the theologians held on firmly to the quintessence as justifying the idea of the higher regions of the heavens as the "place" in which divine love and harmony begin. On the other, Copernicus, without ever drawing the full implications from his theory, had shown how the fixed stars lay at a far greater distance from earth than had previously been thought. Discussion immediately started as to what matter filled these enormous distances. In France Jean Pena, although refusing ultimately to endorse the Copernican astronomy, suggested in his study of optics that a celestial quintessence would cause refraction of light as it passed from celestial matter into sublunar matter, which his observations failed to corroborate.

---

other. To say that all things will be one, if there is nothing besides what is itself, is absurd. For who understands 'what is itself' to be anything but some particular thing? But if this is so, there is still nothing to prevent there being many beings, as has been said. It is, then, clearly impossible for what is to be one in this sense (Aristotle 1984, 1:319).

8. See *On Generation and Corruption*, ibid., 1:512–55, and *Generation of Animals*, ibid., 1111–1218.

9. For Aristotle's matter theory in these terms, see *Physics*, bk. III, 5–8, ibid., 347–54.

Tycho Brahe's more accurate observations would later prove Pena's conclusion to be not altogether correct.[10] Nevertheless, Pena's idea was eagerly picked up by those whose observations of the new comets suggested that they lay above the sphere of the moon, thus contradicting the idea of the eternal unchangeable harmony of the celestial spheres as they revolved in the higher regions of the heavens. It was in the context of this post-Copernican discussion that the renaissance revival of Epicurean philosophy passed from the moral sphere, where it had already become a point of lively discussion on the part of the new humanists, to the scientific sphere.[11] For it had been against the Aristotelian cosmology and the Aristotelian doctrine of matter that Epicurus had reproposed the atomistic theory as valid throughout an infinite space, claiming that our sensible representation of things fails to conform to the ultimate nature of matter, in which the minimum atoms are hidden in subtle and invisible aggregations.[12]

Bruno's debt to Epicurus, primarily through the poem of Lucretius, is made immediately clear in the Frankfurt trilogy by the choice of form: the communication of scientific doctrine in Latin verse. Bruno at once distances himself from Lucretius, however, by alternating passages in verse, which communicate the core of his doctrine, with passages in prose, which explicate and comment on the doctrine itself. This formal distancing corresponds to a conceptual distancing regarding Lucretian atomic doctrine, for Bruno's atomism is, in many respects, at considerable variance with that of Epicurus. All the *loci classici* of the ancient discussion of atomism mentioned above were well known to Bruno, who develops a theory of matter which incorporates a variety of the solutions proposed by classical antiquity rather than adhering to one specifically indicated atomic doctrine. This eclectic stance does not mean that he had any particular desire to mediate between the various and conflicting doctrines in the classical texts. Rather, as Kuhn, in my opinion, correctly surmised, the nature of the atomism developed by Bruno is strictly and rigorously related to his concept of universal infinity in the cosmological sphere. The new theory of matter becomes the foundation of Bruno's infinitism and cannot correctly be understood apart from it.

10. See Pena 1557, f. aa 2v–3r.

11. For early renaissance Epicureanism, the essential study is Garin 1961. See also Copenhaver 1992:198–200. Renaissance discussion of the Epicurean philosophy started in 1417 with the discovery by Poggio Bracciolini of a text of Lucretius's lost *De rerum natura*.

12. See Lucretius 1910, bk. I.

The terms in which Bruno will introduce his atomism into his infinite universe are clearly outlined by his philosophical persona Theophilus in the fifth dialogue of *On the Cause, Principle and One*, the work that immediately follows *The Ash Wednesday Supper*. Once again Bruno insists on going back to the origins of philosophy in Pythagoras, "who did not fear death, but expected only to change." Such change, however, as all true philosophers understand, is only alteration, and not generation or corruption of some basic substance. Real wisdom is possessed only by those who understand that truth, the one, and being are the same. Aristotle did not possess this wisdom because he failed to discover the unity of being.[13] Later on Theophilus, echoing Cusanus, explains this unity in geometrical terms with reference to the identity between the infinite line and the infinite circle. Just as all possible infinite lines start from a point and stretch out to infinity, so all infinite lines can be identified as infinite circles. The maximum and the minimum thus converge into the same unity of being, while at the same time in the dialectic between maximum and minimum all other contraries are resolved into a point of indifference. In the later *De triplici minimo*, Bruno will confirm these geometrical bases of his atomistic infinitism.[14] His infinite universe conforms to Euclidean geometry, which can thus be used as a proper instrument to explicate its forms in space. He will nevertheless argue strenuously against Aristotle's doctrine of the infinite divisibility of matter—minimum parts added to other minimum parts constitute finite bodies—so there cannot be infinite divisibility of matter, either actually or potentially. If there were, finite bodies could not exist.

In the final dialogue of *The Infinite Universe and Worlds*, Bruno is concerned to qualify the void that fills the interstices between the atoms in his now infinite space. He defines it as an "ether" that "contains and penetrates everything": an ether within which everything moves, lives, and vegetates.[15] In the later *De triplici minimo* Bruno sometimes substitutes the word "air" for ether and at times uses the term "vacuum." He makes clear, however, that he accepts the Aristotelian aversion to the vacuum in its strict definition of empty space. Bruno's atomism thus follows neither Democritus nor Epicurus, both of whom thought of an infinite universe filled by atoms interspersed with empty space. Bruno's ether, however, is

13. Bruno 1958:324–25.

14. Ibid., 335–36, and *De triplici minimo*, bk. I, chap. 4, in Bruno 1879–91, I, iii:144–49.

15. Bruno 1958:528. On Bruno's ether see Mendoza 1995:243, n. 3.

not itself atomic in composition. It can perhaps most properly be understood as a substratum of the infinite substance formed by the infinite number of atoms.

All atoms, in Bruno's philosophy, are circular: another factor that differentiates him from Epicurus, who thought of a finite but undefined number of basic atomic shapes, as well as from Democritus, who believed in an infinitely great variety as well as number of atoms.[16] Although taking over a neo-Platonic theme here, Bruno sees in the form of the perfect circle, conceivable only with the reason, above all the means of establishing the identity between the maximum and minimum spheres of being: an identity which is the concept that inseparably links his atomism with his cosmology.[17] Because of his insistence on the circularity of atoms, Bruno is obliged to treat at some length the problem of the difference between the minimum forms of particular species and the absolute minimum or atom. Bruno insists that in the natural sphere all things can be reduced to their own minimum: that is, the minimum cow is always a cow, just as the minimum triangle is always a triangle. Such minimums, however, he defines as "sensible minimums," strictly related to the species of which they are minimums and subject to addition and subtraction.[18] They are not to be confused with absolute minimums or atoms, which are monads or three-dimensional extensions of geometrical points, and whose criteria of combination can be defined as a subtle process of aggregation according to the infinite number of forms possible within the eternal process of vicissitude.

The Aristotelian objection to the atomic theory, developed in books V and VI of the *Physics*, had rested on the argument that since indivisibles have no parts, they must be in contact with one another as whole with whole and so cannot form a continuum. Bruno replies to this objection by proposing the distinction between the concept of atoms as minimum entities and what he calls their *termini*. Accepting the Aristotelian argument that atoms, being indivisible, have no parts, Bruno nevertheless claims that they have limits which touch the limits of other atoms separated by an indivisible distance. Thus it is the indivisible, impenetrable

16. Aristotle often refers to the terms of Democritus's atomism in critical terms. See *Physics*, bk. I, chap. 4, 188a19–26. For the shapes of Epicurean atoms, see Lucretius 1910, bk. I, vv. 479–500.

17. For the necessary circularity of Bruno's atoms, see *De triplici minimo*, bk. I, chap. 12, in Bruno 1879–91, I, iii:177–80.

18. For the minimums of species see *De triplici minimum*, bk. I, chap. 10, in Bruno 1879–91, I, iii:171–74.

atoms, touching one another at indivisible distances, which make up all the forms in the infinite whole.[19]

The question thus arises as to what forces effect the agglomeration of the atoms in space, a problem which both Democritus's vortices and Epicurus's falling showers of atoms solved in terms of largely casual meetings and formations. Avoiding this casual materialism, Bruno introduces the idea of power or energy into his indivisible atoms. Far from being merely minimal spatial entities, combining through chance collisions, as in the Epicurean scheme, Bruno's atoms are seen as each invested with the total power of the infinite cause. In geometrical terms, each atom is the center of the infinite sphere whose center is everywhere and circumference nowhere. This center, however, is far more than a mere point in space. It is a nucleus whose expansive power "explicates admirably" all the sensible forces present in the infinite whole. Thus "the minimum contains more energy than any corporeal mass" to which it has given life by its aggregation with other minimums.[20] Explicator within the infinite universe of the power and intelligence of the divine principle and cause, Bruno's atom is, as it has been appropriately named by Michel, "centre de vie." It is ultimately in the atom that Bruno founds his idea of an intelligent universe governed by intelligible laws.

Bruno refers at times to the powers that vivify the atom as "soul." He uses this traditional theological term to suggest that his atomism can offer a valid argument for our immortality. For the soul contained in the indivisible atom is to be conceived of as its principle of immortality, which orders, agglomerates, disaggregates, vivifies, moves, and weaves together the various transient bodies that make up the mobile universe of becoming. This form-making activity of soul should not be considered an accidental giver of forms, as Aristotle and Galen conceived it, inferior almost to the substantial forms of each particular body to which form is given. Rather, again following Pythagoras, the total amount of soul within each transient body predisposes itself, during the duration of that body, to continue its life in another corporeal form at the moment of dissolution.[21] Thus aggregations are not compounded by casual collisions of the atoms but by the action of the element of soul, intelligence, or energy contained in each of them. Compound entities, including human beings, are not

19. Aristotle's arguments against atomism in these terms are in *Physics*, bk. VI, 231a18–231b17. Bruno's reply is in *De triplici minimo*, bk. I, chap. VII. See Bruno 1879–91, I, iii:154–62.

20. *De triplici minimo*, bk. I, chap. 4, in Bruno 1879–91, I, iii:144–49.

21. *De triplici minimo*, bk. I, chap. 3, in Bruno 1879–91, I, iii:141–43.

formed by a process of mere addition of numerous minimum parts but by an internal principle of intellectual energy, or mind, which governs the criteria of unification. All compounds, including human ones, are nevertheless unstable: "Is it not true that matter varies in time without rest, assuming ever renewed forms once the preceding ones have vanished? Is your own corporeal substance the same as it was a short while ago, either in its parts or in the whole? Are the blood, the flesh, the bones of the adolescent the same as those of the child? And both different from those of the man?"[22] In spite of these Humean doubts about personal identity, however, Bruno continues by claiming that the principle which holds the unstable aggregate together maintains an essential identity until such time as the aggregation collapses and the minimum parts return into the infinite flow of the primal substance. Even then the principle of unity, or soul, survives the collapse, wandering through the vicissitudes of matter in search of another group of minimums which it recognizes as agreeable and into which it inserts itself as if into a new skin.[23]

There are clearly contradictory and unsatisfactory aspects to Bruno's arguments for a Pythagorean metempsychosis, which he justifies explicitly with reference to "the saintly words of the philosopher of Samos." The problem is complicated for Bruno because his atoms are identical. Individual entities thus derive their particular character or form entirely in terms of the principle of aggregation, and not of a basic shape or character of particular atomic particles. It is not clear, however, how this element of energy, mind, or soul within each minimum part itself agglomerates into a unified principle governing the whole. Nor is Bruno able to specify how the individual soul is able to maintain its identity after the collapse of the material atoms back into the primordial substance. Does the individual soul also collapse back into a universal or world soul, losing all its specific qualities? Surely not, for it would be impossible then to speak, as Bruno does, of the immortality of the soul except in entirely impersonal terms. How, then, does the single soul formation journey from its previous material agglomeration to the following one? Given that Bruno specifies that matter and form cannot be conceived of separately from one another, how do particles of soul manage to survive the collapse of the body of which they are such an integral part? Do they remain at-

22. Ibid.; my translation. This chapter, which aims to demonstrate that "death does not involve corporeal substance and even less the soul," is essential for an understanding of Bruno's atomism.

23. For Bruno's sense of the instability of individual things, see Badaloni 1994, who, however, does not pursue the question in terms of Bruno's atomism.

tached to some of the material atoms involved in the collapse? Is there a time lapse between the collapse of one formation and the establishment of another? One might, perhaps, envisage a flash of divine light in which, instantaneously, the departure from one collapsed agglomeration and the formation of a new one take place, rather like the teleportation machine of modern science fiction.[24] But this would surely be to take soul too much out of the process of vicissitude of the eternal substance in time, and Bruno speaks only of a "rapid" process of change. Moreover, he oscillates uncertainly between a traditional theological concept of an individual soul, able to exist entirely separate from material agglomerations of atoms and to dominate them, and a more materialistic Epicurean concept which sees the soul as "following" the course and vicissitudes of events, subject to a blind fate. Furthermore, it is not clear whether Bruno's audience would have welcomed his Pythagorean idea of their immortality through metempsychosis, which clearly reduces the individuality of each soul within its agglomeration of atoms to an at least partly transient, although far from casual, identity.

We should remember that Bruno here is almost certainly arguing also against the idea of the mortality of the individual soul put forward in his century by neo-Aristotelian natural philosophers such as Pietro Pompanazzi.[25] This idea seemed to him as negative as the Christian concept of a fully immortal soul subject, after death, to rewards and punishments outside the sphere of the natural universe: an idea that rests on faith in a transcendental God who performs miracles, such as the incarnation of Christ, in an inferior natural universe. Bruno would cautiously endorse this idea of the divinity, "speaking as a Catholic" during his trial. He added, however, that when he proceeded according to philosophical reasoning he was attracted rather to the reasons of Pythagoras, which he judged "probably" true.[26] As a philosopher, he is thus not concerned with arguments proposed by the theologians for whom the new atomism threatened both orthodox accounts of the creation of the world and any but the most symbolic accounts of the Christian Mass or Holy Communion. Nevertheless, Bruno, who had led a monastic life for many years, clearly felt the need to maintain some form of survival after death of an element he calls soul and associates not only with individual atoms but

24. For a discussion in terms of atomic theory and modern computer technology of the same problem that was worrying Bruno, see Penrose 1989, chap. 1.

25. For the renaissance discussion of the immortality of the soul, see the sections "The Organic Soul," by Katherine Park, pp. 464–84, and "The Intellective Soul," by Eckhard Kessler, pp. 485–534, in Schmitt, Skinner, and Kessler, eds., 1988.

26. See doc. xxii, "Circa animas hominum et animalium," in Firpo 1993:284–85.

with the form or identity of composites of atoms, even if he finds difficulties in doing so in completely coherent terms.

It can be argued that Bruno attains his most interesting results in this context where he associates the element of soul in his atoms with the divine intelligence or *mens*. Bruno's philosophy of the mind is often traditional, based on a thorough reading of both classical and medieval sources, as Leen Spruit has illustrated in his book on Bruno's doctrine of knowledge.[27] Rita Sturlese has suggested that the primary influence on Bruno's philosophy of the mind was neo-Platonism, and above all the philosophy of Marsilio Ficino, whose *Theologia platonica*, in particular, Sturlese has shown to have been thoroughly ransacked by Bruno.[28] The Platonic, and even more the Plotinian idea of an ascent of the *mens* through grades of knowing toward fragmentary glimpses of the divine intelligence, remained metaphorically in Bruno, even after his infinitistic cosmology and his atomic theory of matter had destroyed the various grades of being on which the neo-Platonic concept of the gradual ascent of the soul toward God had rested.[29] It is important, however, not to forget the contradiction between this traditional idea of an ascent through the spheres of knowledge and an atomic philosophy of matter in an infinite universe which posited the idea of mind in the atomic minimum. The idea of "ascent" in Bruno thus becomes inevitably metaphorical insofar as it refers to the sphere of being. The achievement of knowledge of the divine is more a process of penetration by the intellect into the innermost recesses of being, at the heart of which lies the elusive atomic minimum or monad.

One is tempted to use the traditional phrase "the heart of the matter." And indeed it would be appropriate in this context. For Bruno frequently posits the "place" in which the intelligent principle or "soul" in humans and animals resides as the heart rather than the brain. We do not know whether Bruno was in contact in England with scholars already working on the circulation of the blood, which, in the early seventeenth century would be attributed to William Harvey. Some, however, claimed that Walter Warner preceded Harvey in this discovery; and Bruno's work was being read very seriously by the scientists of the household of the ninth earl of Northumberland, of which Warner was a member. Bruno's *De immenso*, which was in the Northumberland library, contains some often quoted passages on the primacy of circular motion which include the idea

27. See Spruit 1988.
28. See Sturlese 1994.
29. For Bruno and Plotino, see Beierwalters 1973/1983.

of the circulation of the blood through the heart.[30] The subject has been studied by Walter Pagel and Jean Jacquot and need not be elaborated on here.[31] Of note, however, is the importance of the idea of an organic life form and growth which underlies Bruno's animated conception of atomism. Operating from within, the primal seeds or atoms contain the energies that cause life to pulsate and evolve throughout the universal whole.

One of the most remarkable pages of *De triplici minimo* points out the apparently irrational and inexplicable behavior of at least some atoms. Here Bruno starts out by treating what he calls the "privative" properties of atoms. As well as being active centers of energy and will, some atoms can also be seen as what today we would call antiparticles: the points of minimum pain, minimum color, minimum sweetness, and so on. Furthermore, although all atoms must be considered incorruptible and immortal, there are some that appear to have less essential properties, and these atoms at times disappear completely. It would be improper, Bruno thinks, to conceive of these atoms as being generated or being subject to corruption. All it is possible to say about them is that sometimes they are present and that then for a space of time they are not.[32] Bruno does not say how he got this remarkable thought. He does not relate it explicitly to occult doctrine or magic but rather to the accidental aspects of the qualities he is discussing, such as color and sweetness. It seems, then, that Bruno is already anticipating here a doctrine of primary and secondary qualities such as Galileo proposed in *Il saggiatore*.[33] Galileo, like Democritus before him, will explain secondary qualities as present only in the mind of the observer, whereas Bruno seems to be thinking of them as some secondary form of atomic entity which has the property of appearing and disappearing. The doctrine is only hinted at and not fully worked out; for if atoms are to be conceived of as eternal and incorruptible, it is clear that they cannot properly be said to disappear. Nevertheless, the page is interesting, and the commentator today may be tempted to relate these aspects of Bruno's atomic treatise to modern developments in particle physics, with its negative energy states and elusive quarks.

30. For the Northumberland collection of Bruno texts, see Gatti 1989:35–48.

31. See Pagel 1951 and Jacquot 1974. It is interesting to find that Samuel Taylor Coleridge copied out this passage of Bruno's *De immenso*, bk. VI, into his notebook in April 1801. See Coleridge, *Notebooks*, ed. K. Coburn, vol. 1 (New York, 1957), 927–29. On Coleridge's reading of Bruno, see Gatti 1996.

32. These pages on the privative properties of atoms are in *De triplici minimo*, bk. II, chap. 6. See Bruno 1879–91, I, iii:208–11.

33. See Galilei 1890–1909, 6:348–50.

Another notable aspect of Bruno's atomic treatise, of particular interest to the modern reader, is his awareness that the inquiry into the minimal components of matter may necessitate using more than one mode of explanation, an idea that questions the Aristotelian principle of noncontradiction on which the classical mechanical science would be so largely based. For example, Bruno refuses to let his corpuscular theory of matter exclude an explanation of bodies according to the traditional concept of distinct elements. In the Italian dialogues, although adumbrating an atomic theory of matter, he still refers to the four elements (air, fire, water, and earth) which Aristotle had inherited from Empedocles. Later on, in *De triplici minimo*, Bruno attempts to clarify his theory by positing three distinct elements, stressing that he does not believe that primary matter is made up of a single substance. His three elements are now light, a humid quality, and a dry quality.[34] It appears from his explanation that the humid quality is to be associated with the air or ether that permeates the infinite universe. The dry quality is associated with the atoms themselves, which are to be found in the substratum of the air. Light seems to constitute what Bruno calls an "indivisible vital force" which permeates the whole, unaffected by place or time. Light thus seems here to have taken the place Bruno accords at other times to the more traditional concept of the world soul. It is not clear whether light, in this connotation, is to be considered as atomic. In any case, the atom is to be considered the "principle" of quantity and measure and the ultimate substance of matter, rather than a piece of matter itself. For any "piece" of matter will already be composed of some agglomeration of atoms, as well as of some combination of the other original elements. Thus the elemental and the atomic theory tend to appear as two complementary explanations of the phenomenon of material forms, rather as modern physics contemplates both a particle and a wave theory of light.[35]

34. For this particularly important aspect of Bruno's mature atomism, see *De triplici minimo*, bk. I, chap. 9, in Bruno 1879–91, I, iii: 169–71.

35. Another way of approaching the question would be to see such double forms of explanation as related to the difficulty of abandoning Aristotelian terminology together with Aristotelian physics and theory of matter. This line of inquiry is developed in Baldini 1977, where the term "element" is considered as one of the cruxes of this situation of terminological transformation. For the four elements in Aristotelian physics are the constitutive elements of prime matter, which, considered ontologically, is not definable in mathematical terms. The process of abandoning the ontological implications held by the term in Aristotelian physics would only be gradual and, in Baldini's opinion, the new *corpuscolarismo* would develop from a philosophy of nature into a series of analytically defined and limited notions operating in the fields of chemistry and microphysics only in the eighteenth century.

Frances Yates, approaching Bruno's atomism through her consideration of his Hermeticism, made the point that he introduced the atoms into his discussion of *spiritus* in the *De magia*. What she calls in a footnote "Bruno's introduction of Lucretian atoms into his natural magic" interested her enough to dedicate a few lines to Bruno's atomic treatise. It is questionable whether she was right in assuming that it had little to offer to her thesis.[36] Bruno treats the phenomenon of light at some length, for the Hermetic cosmologists of his century believed that divine light filled up the infinite post-Copernican universe that lay outside the sphere of the fixed stars. Bruno repudiated this solution as preserving the fundamental Aristotelian dualism between an impure material universe and a purer sphere of material being beyond it. Nevertheless, he will praise Palingenius in the final *De immenso* for his definition of divine light as a special principle to be distinguished from the light reaching us from individual sources such as the single suns.[37]

Divine light is the absolute substance considered primordial by Moses in the book of Genesis, by Hermes Trismegistus, and by other Chaldeans and Egyptians. Bruno believes it is diffused throughout the infinite universe, and not just above the sphere of the fixed stars. Divine light is the eternal principle of life, without which neither the simple substance nor composites could exist. Light thus seems to lie at the interface between the divinity in its aspect of *complicans* and the same divinity in its explication in the infinite universe: an interface which, like the circumference of the infinite circle, should be considered as everywhere and nowhere. As a consequence of this special status, the source of divine light lies beyond the apprehension of the intellect and can never be seen. All that can be discerned is its diffusion throughout the infinite whole. Because of this absolute quality of the primordial light and the impossibility of discerning its ultimate source, Bruno posits the necessary relativity of any apprehension of light in the material universe itself: "the point from which the ray of light moves can only be seen through a medium, accidentally, that is to say in relation to something else." This light that reaches us from some specific source such as the sun or, by reflection, the moon, however, should not be confused with light in its absolute

36. *De triplici minimo*, bk. II, chap. 4, explicitly states that the amalgam of soul and body and the cults and magical contacts between the living and the dead will be treated at greater length in a later work on natural magic.

37. The attentive treatment of the thought of Palingenius, who "roused himself from the dream of a finite universe, but started to dream of an infinite light beyond the world and fell into a thousand other vain speculations," is in *De immenso*, bk. VIII, chap. 4, in Bruno 1979–91, I, ii:295–99.

sense, which is unique and unrelated to any discernible source in the material world.[38] Although not wishing to read into this page of Bruno's *De triplici minimo* more than would be historically correct, we may speculate that in some future incarnation in our own century the Nolan philosopher might have been interested to hear of Einstein's relativity theories and of his conviction of the absolute value of the speed of light.

This brief outline of Bruno's development of his atomic theory demonstrates that it follows no one of the ancient theories reaching him from classical antiquity. Rather, it is an amalgam that maintains characteristics of different and often divergent doctrines. For example, the component of "soul" present in his atoms, as well as deriving from neo-Platonic doctrine, recalls Aristotle's seeds rather than the material atoms of Democritus or Lucretius. The divergences from neo-Aristotelian doctrine, however, must be emphasized. Bruno's atoms are not limited to the animal or vegetable kingdoms or to the purely sublunar sphere. The Epicurean element is most evident in the close relationship established between the infinite dimensions of space and the atom as its invisible minimum component. Bruno also expresses approval of Epicurus's idea, mediated through Lucretius, of our visual images of things as tenuous simulacri, underneath which the minimum quantities, or atoms, remain hidden, outside the sphere of sensible vision.[39] But Bruno repudiated the Democretean and Epicurean ideas on the sizes and shapes of the atoms, as well as their propositions about the ways in which the atoms collided and coalesced. These materialistic atomic theories would be more useful in the coming century as instruments for explaining physical and chemical phenomena and change in an increasingly mechanistic universe. Bruno's more vitalistic atom, with its internal form-making nucleus of light, energy, or soul, however, would find some enthusiastic adherents even in the culture of his times; for although unreconcilable with orthodox Christian doctrine, it retained an essential life-giving creativity and permitted reference to a supreme, divine intelligence.

Bruno does not appear to have developed his atomic theory of matter in relation to the explanation of any particular physical or chemical phenomena. He related his atomism constantly, and almost exclusively, to his infinite cosmology, of which it became the principal foundation and justification. Nevertheless, his vitalistic concept of matter would arouse much interest later on, in the nineteenth century, when it became incor-

38. This particularly interesting page of Bruno's atomic treatise is to be found in *De triplici minimo*, bk. I, chap. 9.

39. Ibid. For the passage in *De rerum natura*, see Lucretius 1910, bk. I, vv. 268–70.

porated into the natural philosophies of the romantic period by Schelling in Germany and Coleridge in England. Furthermore, in the second half of the nineteenth century, several commentators would consider Bruno one of the principal forerunners of an idea of natural evolution.[40] In our own times, it is interesting to think of him in reference to the newest biological discoveries such as DNA.[41] For Bruno claimed already in *The Ash Wednesday Supper*, before he had developed an atomic theory of matter in any detail, that the key—the code—to the secrets of being were to be looked for from within the world:

> And so we are spurred on to discover the infinite effect of the infinite cause, the true and living expression of infinite energy; and our doctrine claims that the divinity should not be searched for outside ourselves, given that we have it with us, we have it within, actually further inside than we are in ourselves; and in the same way, those from other worlds should not search for it in us, given that the moon is to us in the sky as we are to the moon.[42]

The relativization of every point of reference within a now infinite whole deprives any point of reference of absolute value while at the same time suggesting that the individual point of being—the individual body and mind together with the particular world in which such a mind-body complex lives—become the proper subject of study for future generations. Rather than using his atomistic theory in the study of any particular natural phenomenon, Bruno should thus be seen as among those who were using a renewed atomism to lay the metaphysical foundations for a new science.[43] The problem then becomes what intellectual instruments should be proposed as the most adequate means of pursuing the new inquiries.

40. See the relevant pages in Osborn 1894.

41. See Gatti, ed., 1996:47. Also Mendoza 1995.

42. See Bruno 1958:34; my translation.

43. Burtt 1925, which still contains much of interest for this type of inquiry, mistakenly dedicates only a few words to Bruno and his Italian contemporaries such as Telesio and Patrizi, jumping from Copernicus to Kepler, Tycho Brahe, and Galileo as if the second part of the sixteenth century had little to offer to his thesis.

9 Epistemology I: Bruno's Mathematics

Bruno's atomism, as we have seen, was an essential part of an ontology. It was not presented in instrumental terms as a hypothesis useful for investigating precise physical phenomena. Rather, the atom for Bruno was the necessary opposite pole of his infinite universe. When considered under the aspect of the maximum, being is infinite in spatial extension and eternal in time: conceptually it corresponds to an infinite circle whose circumference is everywhere and whose center is nowhere. Considered under the aspect of the minimum it is the atom, the minimum sphere or point of contraction of the energy of the infinite whole. Bruno's epistemology is strictly linked to this ontology. It may be termed a crisis epistemology; for the two fundamental terms of being, the infinitely large and the minimally small, are not clear and measurable entities to the finite human mind. This problematical epistemological stance clearly separates Bruno, at the very beginnings of the modern scientific revolution, from the optimistic founders of the new science such as Galileo and Descartes.[1] Neither the clear and certain intuitions of Descartes nor the mathematical certainties of Galileo, which "knot" logical reasoning into certain and undubitable truths about the natural world, were available to Bruno as means of knowing with certainty the infinite vicissitudes of infinite being.

At times, however, Bruno could sound a more optimistic note. The blind men who arrive on the shores of the Thames under the dark influence of the spells of Circe are given new sight. They look into the two

1. For Bruno's problematical epistemology, see Spruit 1988.

eyes of the English nymph, symbolizing some form of dualism which the *Heroici furori* suggests is a reason/imagination or mind/body dualism. The nymph herself, whose eyes operate the all-important illumination, represents the principle of unity which composes such contraries into a whole.[2] In chapter 5 of book I of his later mathematical treatise *De triplici minimo*, where he claims to be treating the practical application of his inquiry, Bruno defines his sense of the mind's illumination in a significantly different way: "If to the perfection of the human mind, which the fathers of knowledge strive to obtain, one also adds empirical practice, such a light is irradiated from the mind to the senses that they are able to ascend the steps that lead to the highest places, passing beyond the crossroads of the intermediate region."[3] The crossroads here clearly refer to the Pythagorean Y, the conceptual space of moral and intellectual choices beyond which lie "the highest places" obtained through a reconciliation of theory and practice. Alone now in the natural universe, the mind can and must attempt to reduce to measure and intellectual order the infinite vicissitudes of the infinite number of bodies in space.

The traditional Aristotelian logic, which Bruno mercilessly derides in *The Ash Wednesday Supper* as concerned largely with verbal definitions, is clearly no longer sufficient for such a daunting task. Bruno does not want to abandon Aristotelian logic altogether, but he proposes for it a use largely limited to confounding his opponents with their own mental tools.[4] That is, he reduces it to a topics and a rhetoric, useful in debate. Investigation of natural phenomena will require other tools, especially investigation into the existence of those minimum parts and particles which Aristotle himself so strenuously denied. The mathematical doctrine Bruno developed in *De triplici minimo* cannot, in my opinion, be properly understood unless it is remembered that its purpose is to comprehend a particle physics that conceives of the minimum as much more than a simple point in space.

Almost all but the most recent comment on the *De triplici minimo* has been characterized by a sense of frustration and disappointment. At the end of the nineteenth century Felice Tocco expressed intense irritation

2. It is interesting to consider how far, and in what terms, Bruno participated in the kind of mind-body dualism discussed in Rorty 1980.

3. See Bruno 1879–91, I, iii:149; my translation.

4. This method of confounding the opposition by using their own methods of reasoning to demonstrate, through use of the *reductio ad absurdum*, the "imbecility" of their arguments is proposed on the opening page of the second dialogue of Bruno's *Ash Wednesday Supper*: ". . . per desiderio, che tiene, di mostrar la imbecillità di contrari pareri per i medesimi principii, co' quali pensano esser confirmati" (1958:50).

when he found Bruno "mixing" mathematical reasoning with mythological figures and names.[5] In our own century, the pioneering and still important attempt of Xenia Atanassievitch to provide a coherent, chapter-by-chapter reading of Bruno's difficult text was also vitiated by a tendency to praise the rationalistic passages of familiar mathematical doctrine while disregarding as unreadable, puerile, and even morally irresponsible those large parts of *De triplici minimo* which cannot be understood through a classical mathematical approach.[6] Frances Yates, with her usual acumen, suggested that the problem might be that Bruno was not developing a mathematics in the classical sense, but something which she called "mathesis." Yates, however, rather incautiously lifted the term "divine mathesis" out of an altogether different work of Bruno's, the early neo-Platonic work on memory called *Thirty Seals*, and associated it, as Bruno does in that work, with the other three "guides to religion": love, art, and magic. She made no attempt to read *De triplici minimo* as a work of "mathesis," extracting from it only the three complex emblematic figures, the temples of Apollo, Minerva, and Venus (considered among the "puerile" pages by Atanassievitch) whose meaning she attempted to define in terms of magical doctrine.[7] Yates believes the term *mathesis* pertains to medieval astrological tradition. She fails to mention that it was still being used by Descartes, whose "universal mathesis" was rather a meta-mathematics: "a certain general science" which explains everything about order and measure.[8]

This uncertain and unsatisfactory critical tradition of reading Bruno's *De triplici minimo* has at last started to be overcome. Carlo Monti's balanced and illuminating reading of this text in the introduction to his translation into Italian of the Frankfurt trilogy attempts to place it correctly in the context of Bruno's atomism and his infinite, animated universe, as the intellectual tool with which he endeavors to understand his new cosmology. In two major studies dedicated to Bruno's mathematics

5. Tocco 1889.

6. Atanassievitch 1923/1972.

7. See Yates 1964:296–97 and 319–21.

8. Actually, Yates had read the relevant passage of Rule IV of Descartes's *Regulae, or Rules for the Direction of the Mind*; see Decartes 1961:16–18. Her copy of this text, held in the library of the Warburg Institute in London, is even underlined at the relevant passage (p. 16), where she has written in the margin "ancient truth." Her failure to mention this Cartesian page when she discusses the term *mathesis* (1964:296–98) may be due to the unsatisfactory nature of the translation she was using, which renders Descartes's universal *mathesis* as "pure mathematics," substantially distorting the meaning he gives to the term. For Descartes's *mathesis universalis*, see Clarke 1982:166–80. On p. 453, however, Yates herself refers to the Cartesian idea of *mathesis*.

as a whole, Giovanni Aquilecchia has contributed new elements to a critical understanding of this subject.[9] First, through a detailed investigation of the sixteenth-century background of reading and influences behind Bruno's mathematical doctrine, he has shown how it participates in many aspects common to renaissance discussion of the discipline, deserving to be considered seriously as an essential moment of Bruno's philosophical investigation. Second, his consideration of Bruno's pamphlets and successive comments dedicated to the military and geometrical compass invented by Fabrizio Mordente, in the light of an interest in the same subject on the part of Galileo, indicates that Bruno's discussion of mathematical instruments was far more closely linked than had previously been suspected to an ongoing debate in the culture of his time.

Following a more epistemological approach, Luciana de Bernart has claimed that Bruno took over the concept of mathematics as a language of abstract logical "signs" (*entia rationis*) developed in the fifteenth century by Cardinal Cusanus in an attempt to apply those signs as verification of a universe that moves between the two numerical as well as actual extremes of the minimum monad or point and the infinitely large.[10] In a similar, although even more speculative approach, Angelika Bönker-Vallon, in a monograph dedicated to the relation between metaphysics and mathematics in Bruno, argues that Bruno's universe and its first principle are accessible only by mathematical reflection, which in turn is grounded in the first principle of the divine unity. These discussions have established, with considerable clarity and erudition, that Bruno's text is about mathesis in the sense of a metamathematics and that as such it participates in a wider renaissance discussion of the subject; although, as we shall see, it does lead to a somewhat skeptical evaluation of modern mathematical doctrine as Bruno understood it: a skepticism which he had earlier expressed even more forcefully in his *Centum et viginti articuli contra huius tempus mathematicos et philosophos* published in Prague in 1588.[11] Furthermore, some uncertainties in the readings of *De triplici*

9. See Monti in Bruno 1980:11–27 and Aquilecchia 1993a:311–26.

10. Bernart 1986 and 1994.

11. It should be noted that Bruno was by no means alone in his skepticism concerning mathematical doctrine. Such skepticism had already become a well-defined attitude among sixteenth-century humanist thinkers. For the development of a "striking parallel" between Ioannes Ludovicus Vives's distrust of mathematics in bk. 5, "De mathematicis," of his *De disciplinis libri XX*, published in Antwerp in 1531, and the mathematical skepticism of the Portuguese Francisco Sanches, who presented a copy of his book to Bruno, see Elaine Limbrick's introduction to Sanches 1581/1988. These pages also underline the semiskeptical attitude toward mathematics of an admirer of Vives, the Spanish philosopher Fox Morcillo, who in his *De demonstratione, euisque necessitate ac vi* (1556) claimed that it is possible to know something provided that an exact method of inquiry

minimo still remain. These center on two issues: Bruno's doctrine of the triangle with its accompanying rejection of trigonometry, and what is still considered a fundamental inconsistency on Bruno's part on the issue of the squaring of the circle. Also, no fully satisfactory explanation has yet been put forward of the presence in Bruno's mathematical doctrine of mythological elements which seem to pertain to a different sphere of discourse. Hence the special characteristics and limits which Bruno imposes on his concept of mathematics as an intellectual tool with which to define quantitatively the new infinite, atomistic cosmology have yet to be adequately defined.

My own approach to the subject will concentrate on a reading of *De triplici minimo* as the work in which Bruno brought to completion his reflection on the mathematics of his time by linking it inextricably to the definition of his atomism. It seems advisable to approach the text, in the first instance, through the lucid evaluation of it supplied by Bruno himself in the dedicatory letter to the Duke of Brunswick, where he describes it as an erudite and scholarly work in which understanding of the primary elements is obtained. On the other hand, according to Bruno, it is to be seen as an expression of a desire to reach the truth rather than an entirely successful project. The work must be considered as based on observation, while at the same time it examines those things which in us are innate. It takes into consideration simple objects: the elements considered are limits, the minimum, size, founded on the straight line, the angle, and the triangle. The ultimate demonstrations of these elements are the temples of Apollo, Minerva, and Venus, formed by touching, intersecting, and circumscribing circles. In these temples all figures, numbers, and measures are implicit, represented, and explicit in virtue of the definitions, axioms, and theories they embody.

This is Bruno's own evaluation of what he was doing in *De triplici minimo*. It suggests that Frances Yates was right to look for the clue to Bruno's mathematical doctrine in the three emblematic temples of Apollo, Minerva, and Venus, even if she failed to place them in their proper mathematical setting. For they do represent properly mathematical concepts, even if they are apparently far removed from the mathematics that would develop in the seventeenth century based on the idea of a direct relationship between mathematical entities and the natural world. Bruno's emblems, with their memories of the "figures" of ancient mythol-

---

is established. In his later *De naturae philosophia*, published in 1560, Morcillo claimed that geometrical method was the most satisfying and that the definitions, axioms, and hypotheses of the mathematical sciences seemed true and certain to him, for at least they could establish probable opinion. See Sanches 1581/1988:34–35.

ogy, represent the values of mind, intellect, and love. They stretch back to what Frances Yates called "ancient truth": at one level—although not the one that Yates had in mind—they can be seen as nothing more or less than pictorial representations of the elements of Euclidean geometry.[12] Bruno's temples, however, also leap forward to a post-Kantian era that insisted on the innate quality of epistemological discourse in which the experience of perception is formulated and controlled by mental axioms, definitions, and theories. The temples are thus neither abstract entities nor magical seals. They are rather the intellectual coordinates or the measuring devices through which the mind approaches the physical world.

In this sense, it is not entirely inappropriate to consider Bruno's "halls" of Apollo, Minerva, and Venus in relation to the modern concept of quantum mechanics: itself a reply to the dilemma of measuring the elusive realities of minimum particles. In modern quantum mechanics, the ever-receding "object" exists only in terms of the notion of measurement, and a gap opens up between the system being observed and the mind of the observer. That gap is occupied, in quantum theory, by a measuring ensemble or device which is located firmly in the external world. Quantum theory has been described metaphorically as a "black box" that gives useful results without supplying any picture of, or assigning any particular reality to, the objects assumed by most scientists to lie beneath their observations. The idea of producing an ensemble of systems by which, or through which, an observable A is measured, without its being in any way "known" or revealed as an object, is essential to quantum theory. Bruno's temples or halls dedicated to the principal Greek mythological gods can be seen as embryonic examples of such systems. Through them it is necessary to proceed in order to reduce to measurement an elusive chaos of vicissitude composed of invisible particles of matter.

It is no coincidence that these temples immediately precede the fifth and last book of Bruno's work, dedicated to the concept of measure. They contain all the primary Euclidean concepts or "systems" of measure, based as they are on combinations of the circle, the triangle, and the square. They may be connected with the mystical Pythagorean tetractys or number 4, as well as the Euclidean *Elements*, or even with similar signs, which Bruno may not have known, found in early Indian ritual geometry such as the *Sriyantra*, which consists of a fourteen-sided polygonal construction at the center surrounded by eight- and sixteen-petaled lotus flowers surrounded by three circles before an outer boundary con-

12. See Yates 1964:319. Similar seals, as Yates points out, are to be found in Bruno's earlier *Articuli adversos mathematicos*.

taining four "doors" leading to the outer world.[13] However interesting such links may appear historically, however, it is important to note that Bruno himself states explicitly that his temples need no clarification from references external to themselves.

In relating Bruno's mathematics to post-Kantian and even quantum theory concepts of measurement and the mind, I am not attempting to present Bruno as a precursor of philosophies of science or concepts of measurement far more sophisticated than anything that could be imagined in the sixteenth century. I am not claiming that what Bruno had in mind already was quantum mechanics. I am nevertheless convinced that the difficulties in correctly reading and interpreting his mathematical works derive largely from a fundamental discrepancy between his mathematical doctrine and the classical mathematics emerging in his time, based on consecutive processes of reasoning seen as creating a mirror image of the order in the natural world. Bruno had in mind, although in a very embryonic form, a quite different idea of what mathematical doctrine was and how it related to the mechanics of his atomic and infinite universe. It is an idea that can be related to ancient mathematical icons or temples, leading from higher forms of rational certainty into the chaos of the phenomenological world. It can also be closely related to the modern breakdown in the concept of an object and to the proposal in quantum mechanics of measurement as a grid through which a shadowy and "virtual" world of objects can (possibly) be indirectly grasped. It is with such similarities in mind that I shall approach some of the major cruxes of Bruno's mathematical doctrine as he develops it in *De triplici minimo*.

In his mathematical, and indeed also his cosmological works, Bruno seldom makes even the most elementary numerical calculation. It must, I believe, be assumed that he could have done so if he had wished, even if his technical knowledge of mathematics went little beyond textbook level. All commentators have assumed that he had very limited technical preparation in mathematics although such an assumption is based more on a lack of evidence than on any incapacity demonstrated by the texts themselves. It is true that Bruno's concept of number was still apparently Pythagorean, that is, he worked only with what are now known as the "natural" numbers based on the decade. One way to understand this problem is to underline the terms of the strictly philosophical approach to both cosmology and mathematics in Bruno's work. His concern in *De triplici minimo* is with the *possibility* of measuring within an infinite universe composed of an atomic substance, not with the ways in which

13. For the *Sriyantra*, see Barrow 1992:73–81.

such measurements are technically carried out.[14] At the other extreme, as most commentators have stressed, he tended to be unconcerned with and deeply suspicious of all forms of abstract calculatory reasoning which, in Platonic terms, aspired to a purely intellectual truth. We shall see, however, that Bruno was not always consistent in this attitude and that some of his fundamental ideas are expressed in terms of a conviction that some of the mathematical objects he is concerned with derive directly from God.

The possibility of measurement depends on some kind of reference frame according to which one point of reference can be compared numerically or spatially with another. Bruno sees the basic concept of the reference frame as one of limits: the first of the elements he claims to be considering in *De triplici minimo*. The tension that underlies all his comments on mathematics derives from the fact that his infinite cosmology had deprived the universe, and the idea of space, of what for centuries had been considered its finite outer, or, in numerical terms, upper limits. In Bruno's mind, a lower limit, both logically and physically, was absolutely necessary. Without the atom, all matter would collapse into nothingness. Without the monad, numbers would lack any foundation. Without the point, spatial determination would prove to be impossible. Bruno's way of approaching mathematics as a dramatic existential problem rather than an abstract intellectual exercise can, in my opinion, be properly understood only in terms of a crisis caused by a radical "paradigm change" (to use once again that convenient, if perhaps overworked, Kuhnian phrase).[15] For this reason, the first book of *De triplici minimo* is concerned with proving "the existence of the minimum." If such a proof cannot be established, all will be lost. It is perfectly coherent with this approach that, having established the terms of his argument in the first two chapters of book I of *De triplici minimo*, Bruno dedicates his third chapter to the subject of death.

I have considered Bruno's discussion of death and immortality through metempsychosis in the section on atomism. It is a discussion based on an atomic conception of matter, and it exposes Bruno to the accusation of presupposing the very postulate he is seeking to demonstrate: that is, the necessary existence of the minimum. He could presumably have replied that he is only reminding his reader, at this point, that in his view the pos-

14. On this subject, see Bönker-Vallon 1997.

15. This is, of course, another factor linking Bruno's mathematical thought to the crisis that produced the new quantum mechanics in the 1920s. For an account of the dramatic development of that crisis, which involved, among others, Niels Bohr, Max Born, Werner Heisenberg, Wolfgang Pauli, and Albert Einstein, see Miller 1984, chap. 4.

sibility of being and of life itself depends on the existence of the minimum: he is not using these arguments to demonstrate their existence. Bruno's arguments, as they are developed in the remaining part of book I, involve above all a frontal attack on the Peripatetic doctrine of the infinite divisibility of matter. Again the basic concept is one of limits. In the expanding order of matter and of number there can now be no limit, for the maximum is never reached. In the descending order, instead, the minimum must be recognized both materially and conceptually; for minimum particles added to other minimum particles constitute finite bodies, in which they cannot be considered infinite in number. It is at this stage in his argument, as we have seen in his discussion of atomism, that Bruno proposes the important concept of *termini* between the minimum points, a fundamental concept necessary for the minimums to form a continuum. The *termini* are the limiting points of contact between the minimum particles and have no extension in space.[16]

Atanassievitch suggests that if Bruno thinks of his atoms as corresponding to the fundamental unity or monad, which he sees as the basis of all number, his *termini* must correspond to zero. Although Atanassievitch herself draws no conclusion from this suggestion, it seems to me that in that case one way of representing Bruno's universe would be through a system of binary numbers, which is certainly an intriguing possibility given the importance of binary numbering in modern scientific development. It is doubtful, though, whether a binary number system can be seen as anything more than implicit in Bruno's idea of atomic monads limited by *termini*. He never explicitly suggests that his *termini* might have zero value, except perhaps where he associates them with the Democretean vacuum between atoms.[17] As Bruno follows Aristotle in refusing to allow an absolute vacuum in nature, this could be taken to mean that the *termini* do have a number, which would have to correspond to zero. It is worth remembering that one of the first attempts to formulate a system of binary numbers was made by the English mathematician Thomas Harriot, who is not only known to have been reading Bruno but appears to have accepted his infinite atomistic cosmology.[18] It is perhaps no coincidence that Harriot copied out into his manuscript papers, in En-

16. For Bruno's development of this argument and its influence on the atomism of the group of scientists surrounding the ninth earl of Northumberland, see Gatti 1989:57–65.

17. It is interesting to note that not all mathematical systems have contained a concept of zero. The Hindus apparently had a sign for it very early on; but as far as the European tradition is concerned, the Babylonians invented it and later the Arabs introduced it into Italy. The word seems to have arrived through the Venetian dialect, in which the Arab *zefiro* became *zefro* and then *zevero*. See Barrow 1992:81–101.

18. For Harriot's system of binary numbers, see Shirley 1951.

glish, a passage from the sixth chapter of book I of *De triplici minimo*, in which Bruno claims to establish the necessary existence of a minimum quantity, absolutely indivisible: "And yet for a last in decreasing progressions we must needes understand a quantity absolutely indivisible; but multiplicably infinitely infinite till a quantity absolutely inmultiplicable be produced which I may call universally infinite. And in encreasing progressions we must needes understand that for a last there must be a quantity inmultiplicable absolute, but divisible infinitely infinite till that quantity be issued that is absolutely indivisible."[19]

Bruno's almost obsessive concern with the mathematical implications of extreme or "crisis" entities such as the inmultiplicably infinite or the absolutely indivisible projects his speculation toward a remarkably modern sphere, calling to mind the work of Georg Cantor. Bruno's work can, of course, not be compared to the extraordinary sophistication of the formal and calculatory procedures achieved by modern mathematicians. Nevertheless, some of the ideas which Bruno is developing in the light of the problems of comprehension and measurement presented by the infinite universe and the minimum atom resemble, in embryonic form, much thinking in our own century about essentially the same problems. Bruno might well have been sympathetic, for example, to a theorem such as Kurt Gödel's, which proves "the necessary limitations of formalized mathematical reasoning. No matter how comprehensive we think we have been, there will always be some propositions which escape the net."[20] Such an insistence on the advanced nature of Bruno's mathematical speculation may seem surprisingly in contrast with what his detractors have always assumed to be one of the most retrograde and limited aspects of his philosophy, and which even his admirers have generally qualified as unsatisfactory or at least contradictory. It would certainly be a mistake to suggest that no inconsistencies or curious misunderstandings on Bruno's part exist with respect to what mathematics is about. But that may be precisely because he was proceeding in what, for his time, were not very orthodox paths.

The so-called and to some extent rightly called dilemma represented by Bruno's mathematics centers on his surprisingly ferocious repudiation of trigonometry—one of the most rapidly advancing fields of the mathe-

19. In BL Add. MS 6782, f. 363r.

20. This is the definition of the theorem given by Hofstader 1985. For Gödel see also Shanker 1988, which gives the famous 1931 paper read by Gödel at Princeton. An article in this volume by Stephen C. Kleene states that Gödel found difficulty in indicating exactly the sphere in which his results could be generalized. He therefore observed that they should be equally applicable to formal systems of mathematical set theory and to the rules of logic.

matics of his time—and his related refusal to participate in the heated renaissance debates over the squaring of the circle.[21] Even here, however, some curious blind spots exist along with some other surprisingly advanced mathematical intuitions. This subject has been often discussed by the commentators and can be summed up briefly. Bruno's atomic doctrine, and his consequent concept of being as involved in an eternal and infinite process of vicissitude, was linked to a decided affirmation of the traditional concepts of genus and species. The specific physical formations of finite bodies are not achieved by pure chance or casual formations, but by intelligent and intelligible laws imposed by the element of mind, soul, or light which reflects the divine intelligence in the infinite universe. The nature of every finite entity is thus conditioned by the nature of the species to which it belongs, while it belongs to it; and Bruno, in the introductory poem to "The Malcontent" in *The Ash Wednesday Supper*, advises the frog to act as a frog and the fly as a fly, without confounding the natural order.[22] Analogously, he thinks that the formal elements of geometry are defined by their constituting axioms and must not be confounded. Every form must be considered according to its defining characteristics; and the approximations of the trigonometers who attempt to identify the minimum arc with the minimum chord can result only in errors. "Away with the useless tables of sines," cries Bruno, incautiously dispensing with one of the major mathematical achievements of antiquity that was subject to a particularly lively development in the sixteenth century in relation to the increased dimensions of the Copernican universe which Bruno took as his point of departure.[23]

Before attempting to inquire into the consequences of such a remarkable attitude, we must clear up some misunderstandings that are still part of the critical tradition. For, however one may want to comment on the reasons for, or unreason of, Bruno's summary dismissal of trigonometrical tables, once he had made it, he was consistent in maintaining and developing it. In particular, the squaring of the circle is commonly considered an issue on which Bruno expressed diametrically opposing attitudes. We now know that the circle cannot ever be equal to the area of a square because pi is not a rational number, although this explanation would not be given until the nineteenth century. Archimedes had proposed ever closer approximations of the circle with the square, and in the sixteenth century hopes were high that a geometrical solution would be found to a

21. See Aquilecchia 1993a:319–26.
22. Bruno 1958:5.
23. For the development of trigonometry during Bruno's century, see Aquilecchia 1993a:312–13.

problem the ancients had failed to solve satisfactorily. Bruno thought, rightly, that all these attempts were based on a misconception and that a fundamental incommensurability existed between the minimum arc and the minimum chord. His realization that no arc could ever exactly equal its subtended chord, however small it became, appears to have been intuitional, and no effort was made by Bruno to provide a rigorous proof.

Bruno's argument, rather than involving the nature of pi, which he never mentions, takes into account the nature of the gnomon of that portion of the circle constituted by any arc and its subtended chord. A gnomon is any part of a geometrical figure subject to expansion or diminution without changing its basic shape. Bruno is saying, in effect, that the portion of the circle made up by an arc and its subtended chord, however small it may become, can never simply vanish into nonbeing. Bruno thus became convinced that the squaring of the circle was not simply a problem that had not yet been solved but that, in the terms proper to Euclidean geometry, it was insoluble. This conviction, which was correct, became one of Bruno's central mathematical dogmas, suggesting to him that the use of numerical approximation, introduced by Archimedes to allow close approximations between squares and their relative circles, was admissible in the sphere of practice but that attempts to arrive at a properly geometrical identity between the square and the circle were mistaken. Following this line of reasoning, it seems curiously contradictory that Bruno should find the approximations of trigonometrical tables so uncongenial.[24] His attitude here can probably be understood only by realizing that he thought of them as threatening the establishment of ultimate geometrical minimums, which he saw as necessary to the possibility of being. Such a contradiction, however, regards his attitude to trigonometrical tables and not to the squaring of the circle, a subject on which he seems to me to have been remarkably coherent.

Following a lead put forward in the nineteenth century by Felice Tocco, modern commentators nevertheless assume that Bruno contradicted this attitude in several passages: an assumption which I believe to be mistaken. Tocco claimed, in particular, that Bruno "spoke two languages" when he referred to the squaring of the circle, and he pointed for confirmation to a passage in *The Expulsion of the Triumphant Beast* where it appears that Bruno accepts the squaring of the circle.[25] That the passage indicated by Tocco does in fact constitute a contradiction is accepted by

24. Bruno's criticism of trigonometry reaches its climax in bk. III, chap. 9, of *De triplici minimo*.

25. See Tocco 1889. For the pages discussed in the following passage, see Bruno 1958:755–56.

Carlo Monti and even by Aquilecchia; but I believe their acceptance of Tocco is based on a misreading. The problem arises in the context of a reform of the heavens which is being carried out methodically by a repentant (but not too repentant) Jupiter, who is chasing out the traditional and by now corrupt signs of the zodiac from their ancient places in the heavens and installing new virtues in their places. In the course of this massive project for reform, Jupiter has to decide what to do with the sign known as the triangle. His daughter Minerva, who is helping Jupiter in his reform, suggests giving it as a present to Cardinal Cusanus, who had proposed several methods for squaring the circle. One of the methods of squaring the circle referred to by Bruno in this passage may be called "the method of the halfway house," although Minerva, who interprets for Jupiter the argument of Cusanus, actually calls it, in hushed tones, the method of "the place of all places."

What it amounts to is that a square is drawn inside a circle with its extremes touching the circumference, while outside the circle another square is drawn touching the circumference at the center of each side. The solution proposed by Cusanus is that the square that lies exactly halfway between the internal and the external square is equal to the circle: a solution which, looking at the figure, seems quite convincing. Actually it is wrong; and when Bruno wrote his pages on the triangle, the proof of Cusanus's mistake had already been put forward by Regiomontanus (Johannes Müller, 1436–76) in a much discussed sequel to his book on triangles, *De triangulis*, published posthumously in 1533.[26] Bruno develops a similar figure and argument using a half of a square, or a triangle, given that he is discussing the constellation known as the triangle as part of Jupiter's universal reform of the heavens.

The book by Regiomontanus, one of the major mathematicians of the renaissance, seems to have attracted considerable attention in the Wittenburg of Melanchthon. Regiomontanus couched his refutation of this and other arguments put forward by Cusanus to prove the squaring of the circle in the form of a dialogue between two ancients called suggestively Aristophilus and Critias.[27] This was a subtle way of calling on ancient au-

26. For an English translation, see Regiomontanus 1533/1967, which, however, fails to include the all-important sequel. For a comment on Regiomontanus in the culture of his time, see Rose 1975:90–117.

27. This form of proof of the squaring of the circle had actually been put forward in ancient Greece by a Sophist of the fourth century B.C., Bryson, and is discussed by Aristotle in *Sophistical Refutations*, XI, and *Posterior Analytics*, I, 9. Aristotle already rejects this proof, although on explicitly nonmathematical grounds. His argument is that Bryson may possibly have proposed a figure of a square and a circle that correspond to each other, but that even if he has, his proof remains sophistical because it depends on visual intuition

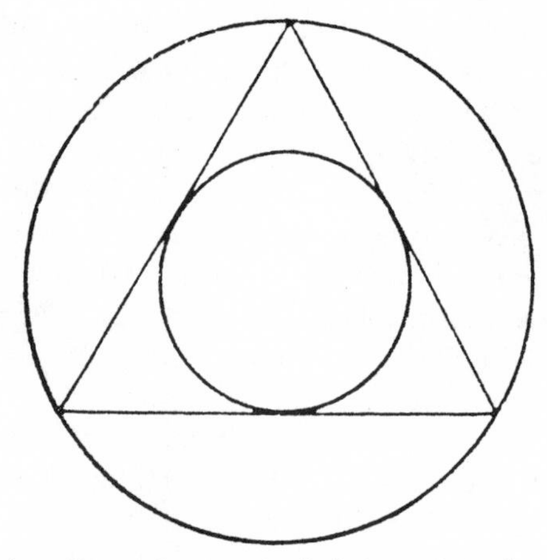

9. Diagram in Giordano Bruno, *Lo spaccio de la bestia trionfante* (1584), dialogue 3. The diagram, which is announced in the text of the first editions, apparently remained unprinted until the nineteenth century.

thority to refute the cardinal, and Bruno (perhaps less of a disciple of Cusanus than he is generally considered) seems to have appreciated all the ironies of the case. His page in *The Expulsion of the Triumphant Beast* must, I believe, be read as satire, bearing in mind that the work as a whole was dedicated to the Protestant Sir Philip Sidney. For the reform being instituted by Jupiter is clearly carried out in the name of an infallible concept of authority which Bruno—perhaps incarnated in the cynical but ultimately helpless Momus—could hardly endorse. This was not the right reform, in Bruno's eyes, nor, of course, in Sidney's either. It failed even to contemplate the new cosmology, which was the only context in which Bruno thought that reform could be carried out with any possibility of success. Jupiter and Minerva's endorsement of Cusanus's arguments, without apparently having been briefed by Mercury that they were already out of date, only underlines the profoundly mistaken nature of their reform as a whole.

Before leaving this page of *The Expulsion* it is necessary to consider briefly the figures presented by Mercury to explain to Jupiter how the squaring of the circle works by the method of the halfway house. The first figure presents no problems, even if the midway circle representing the solution is not actually indicated. The reader, it seems, has to work out for himself where "the place of all places" actually is. Later on, however, Bruno adds two further figures that help to break down Cusanus's argu-

---

and is not argued on grounds proper to geometry. For a discussion of ancient arguments concerning the quadrature of the circle, see Knorr 1986:25–39, 76–86, 153–70.

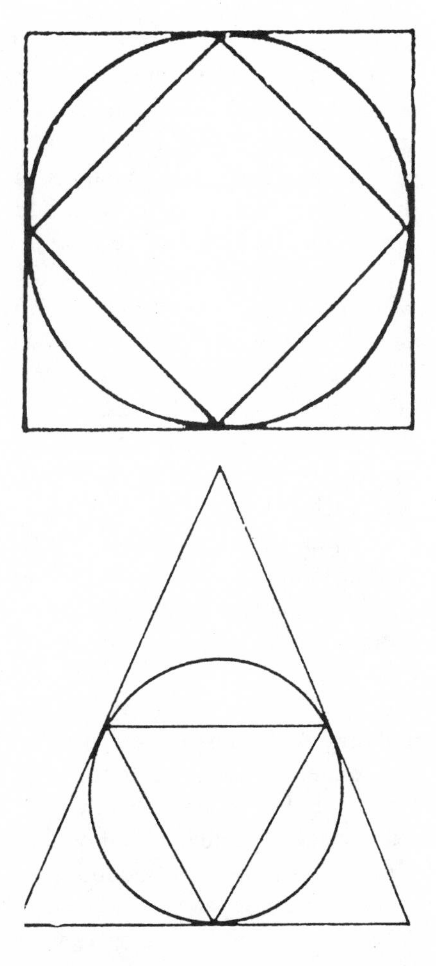

10. Diagrams in Giordano Bruno, *Lo spaccio de la bestia trionfante* (1584), dialogue 3. The diagrams, which are announced in the text of the first editions, apparently remained unprinted until the nineteenth century.

ment. One of these stands an internal triangle on its head; the other tilts an inside square on to one of its corners.

These figures do not establish a halfway house, even visually, for the figures of the triangle and the square become aperiodic. A game seems to be being played here which may remind us of the modern preoccupation with problems of tiling, and, like the modern tiling game, it is also a serious business.[28] Bruno is trying to establish the danger of error inherent, in his view, in the whole exercise of visual approximations. He is not altogether out of tune with Jupiter himself when, in the final remarks on the triangle, the heavenly thunderer brushes aside the subtle arguments

28. For the problems relating to tiling, see Penrose 1989.

of Minerva and Mercury to remind them that the triangle is the fundamental symbol of religious faith. Bruno himself was antitrinitarian and would not have endorsed Jupiter's final evaluation. Nevertheless, he, no less than Jupiter, needed the triangle to maintain its stability. It could only at great peril to the whole structure of the phenomenological world disintegrate into a circle, with which it remained totally incommensurable, in Bruno's view, according to divine disposition.

This leaves us with the apparently puzzling pages in *De triplici minimo* where, outside the context of Jupiter's putative reform, Bruno specifies the terms in which he is prepared to contemplate identifying a triangle with a corresponding circle. At one point (book III, chapter 9) Bruno suggests that instead of considering the circle and the triangle in two-dimensional terms, the square (or, indifferently, the triangle) is drawn on the surface of a sphere. Will the shift to spherical geometry change the nature of the problem? Bruno thinks not. All that will happen is that the projection of an equilateral triangle onto the surface of the sphere will constitute an "analogy" of the parts with the parts that make the case similar to plane geometry. He may be thinking of the projections made by geographers or navigators of areas on the spherical surface of the globe onto charts, which are also based on approximations of small curves of the convex surface with straight lines. So the problem remains the same, and Bruno refrains from elaborating the point in terms of spherical geometry, preferring to return to the problem as it presented itself in the plane. He simply repeats that there will never be direct proportions between an arc and its subtended chord, whatever their size. There is no contradiction here with Bruno's previous position. On the contrary, it is only a way of reasserting it.

Bruno is quite clear in specifying only two senses in which the squaring of the circle may be properly contemplated, both cases lying conceptually beyond the sphere of calculation which may be defined as strictly mathematical. The first case, propounded originally in the fifth and last dialogue of the *De la causa, principio e uno*, and again, in precisely the same terms, in the fourth chapter of book I of *De triplici minimo*, is when the circle being considered is of infinite dimensions. Here Cusanus is called into the picture, for it is from his works that Bruno learns of the infinite as the dimension in which all contraries are resolved, even mathematical ones. Bruno delineates eight modes in which all measure collapses and becomes meaningless once infinite quantities are being considered: in the infinite "there are no differences between the infinite circle and the infinite line, the infinite diameter and the infinite area, the center and wherever else; just as there is no difference in the point, which is nothing other than the minimum circle." In this anomalous sphere of infinite quantities, which Bruno thinks of as outside the field of measurement and mathematics, the circle, the square, and indeed all other forms and dimensions become part of the same infinity.[29]

The only other context in which, in Bruno's opinion, it is permissible to talk of squaring the circle is when the means adopted to identify the two figures are "external" rather than "internal" (book III, chapter 12), by which he means practical rather than logical. Here Bruno claims that attempts to construct identical circles and squares with mathematical instruments can be successful, provided that there is no assumption that the squares so constructed are *mathematically* identical to the circles.[30] Bruno does not appear to express approval here for a practical use of the approximations of trigonometrical tables or calculations, based on the virtual identity of the minimum arc and the minimum chord. Rather, he emphasizes that his is an instrumental approach that requires great technical skill and attention, and presumably adequate measuring instruments as well. It is in this context that we should understand the pages in *De triplici minimo* where Bruno expresses a renewed interest in and esteem for Fabrizio Mordente and his military compass, capable, as Mordente himself expressed it, "of demonstrating the perfection of Geometry . . . by not leaving over any residual measure or weight, while it continues dividing infinitely until it discovers the ultimate single mea-

29. Thomas Harriot's manuscript treatise *De infinitis*, which at times seems to echo Bruno, also insists more than once on the anomalous case represented by infinite quantities, asking, for example, whether it is possible to apply the notions of equality and inequality to infinites. Harriot's treatise is in BL Add. MS 6782, ff. 362–74v. See Gatti 1989:57–68.

30. In his preface to Billingsley's translation of Euclid's *Elements of Geometry*, John Dee, in "The Squaring of the Circle, Mechanically" (ff. cj v–c.ij r) describes the following method:

> . . . because it is well demonstrated, that a Cylinder, whose heith, and Diameter of his base, is aequall to the Diameter of the Sphaere, is Sesqualiter to the same Sphaere (that is, as 3 to 2) to the number of the weight of the Sphaere, adde halfe so much, as it is: and so have you the number of the weight of that Cylinder. Which is also Comprehended of our former Cube: So, that the base of that Cylinder, is a circle described in the Square, which is the base of our cube. But the cube and the Cylinder, being both of one heith, have their Bases in the same proportion, in the which, they are, one to another, in their Massiness or Soliditie. But, before, we have two numbers, expressing their Massiness, Solidities, and Quantities, by waight, wherefore we have the proportion of the Square, to the Circle, inscribed in the same Square. And so we are fallen into the knowledge sensible, and Experimentall of *Archimedes* great Secret.

31. Bruno wrote four dialogues, contained in two volumes, on the invention of his compass by Mordente, whom he had met in Paris: *Dialogi duo de Fabricii Mordentis Salernitani prope divina adinventione ad perfectam cosmimetriae* and *Dialogi Idiota triumphans, De somnii interprettione, Mordentius, De Mordentii circino*, both published in

sure or weight to the untold satisfaction of the intellect."[31] So it is the practical applications of geometry that Bruno is considering, accepting that in the real world there will be nothing but approximations anyway. Such an acceptance, however, is not to be confused with the logical, mathematical problem of the squaring of the circle which Bruno, unlike most of his contemporaries, believed to be intrinsically insoluble and not simply unsolved up to his day and age.[32] This distinction between practical mathematical constructions and logical-mathematical reasoning probably lay behind Bruno's constructions of regular polygons in the Pythagorean *De monade*. Bruno's readiness to accept geometrical approximations in the practical sphere of construction got him into trouble with the more mathematically rigorous Kepler, who took him to task, in *Harmonice Mundi*, for suggesting a possible way of constructing a regular enneagon: that is, a regular nine-sided figure which actually constitutes a mathematical impossibility because its sides are unknowable.[33]

It seems difficult to reconcile this acceptance of practical approximations in geometrical constructions with Bruno's violent repudiation of the trigonometers. Nevertheless, *De triplici minimo* is a frontal attack on trigonometry, which is now considered to be one of the most advanced and promising developments of renaissance mathematics, in the name of one of the most advanced and promising developments in sixteenth-century theory of matter: the renewal of ancient atomism. I must stress that Bruno did not deny the possibility of resolving geometrical figures generally one into another. In chapter 16 of the fifth and last book *On Measure* he describes what he calls a "Key to Transformations," taken, without explicit recognition, from Archimedes.[34] There he indicates schematically how a triangle can be constructed with the same area as any other regular geometrical figure, with the sole exception of the circle. The circle, in Bruno's opinion, is the form both of the infinite universe and of the minimum atoms, and he clearly thought that there was some

Paris in 1586. Both volumes were extremely rare and virtually unknown until they were reissued in 1957. For Bruno's strangely ambivalent relationship with Mordente and his work, see Aquilecchia's introduction to that volume, and 1993a:107–18 and 315–17.

32. Bruno would already have found a similar line of argument in Proclus, who in *In Primum Euclides Elementorum Commentarii* wrote: "Those squaring the circle did not seek (to establish) whether it is possible that there exists a square equal to the circle, but rather thinking that it can be so, they tried to produce a square equal to the circle" (quoted in Knorr 1986:76–78).

33. See Kepler 1997:89–91.

34. For the survival of the doctrines of Archimedes in the Western tradition, see Clagett 1961.

absolutely inviolable propriety to the fact that the circle could not be resolved into other geometrical figures. Such a conviction does not make him any less of a mathematician in this work. From Plato onward, there have always been mathematicians convinced, as a present-day mathematician expresses it, "that mathematical objects have a timeless existence of their own, not dependent on human society nor on particular physical objects."[35] Whether Bruno really was a Platonist is probably doubtful. He identified himself more easily with Pythagoras, who saw number as the immanent foundation of being rather than as absolute and transcendent. Nevertheless, when he considers the particular properties of the circle, Bruno at times does seem to assume a quasi-Platonic stance.

I have tried to show that Bruno's arguments concerning the squaring of the circle cannot be attacked on the grounds of internal contradiction, as they nearly always have been; for he is consistent in his intuition that the squaring of the circle represented an insoluble problem in terms of the mathematics of Euclidean geometry. He can, however, be accused of taking up, overall, an unreasonable attitude to trigonometry based on the belief in the necessity for approximation in the calculatory sphere which (far more an Aristotelian here than he probably would have been willing to admit) he was ready to allow and even encourage in instrumental terms. What Bruno really objected to was probably the confusion of the terms in which the trigonometers justified their approximations. In his *De triangulis*, Regiomontanus had admitted that approximation was not truth, adding that nevertheless he preferred to know "the near truth rather than to neglect it completely, for it is worthwhile not only to reach the goal but also to approach close to it."[36] This idea would have been anathema to Bruno. Truth had nothing to do with it, in his opinion. That could be claimed only when logical reasoning reached a satisfactory conclusion in the terms of the definitions posited; and this was not the case in trigonometry. The only justifying factor had to be the practical necessity for carrying out measurement in an imperfect phenomenological world, which, in Bruno's opinion, should be clearly distinguished from anything called logical truth. Support for Bruno's stand came from two very different works that seem to have supplied the conceptual bases of his mathematical doctrine: one of them a contemporary text which only Bruno appears to have appreciated, and which almost at once seems to have fallen into complete oblivion, and the other the major mathematical work of classical antiquity.

35. See Penrose 1989.
36. Regiomontanus 1533/1967:35.

In 1581, the Parisian printer Gourbin published a mathematical work by Mauritius Bressius titled *Metrices Astronomicae*. Bruno mentions this work three times with clear approval in the third book of *De triplici minimo*, concerned with the discovery of minimum quantities. Giovanni Aquilecchia has established that Bressius and Bruno were closely linked in their intellectual activities in Paris.[37] Not only was Bressius's mathematical work published by the same publisher who produced Bruno's first work to have survived, the *De umbris idearum*, a work on the art of memory, but it is probable that both of them taught at the Collège Royale under the direct patronage of the French king Henri III. None of the commentators, however, seem to have addressed the question of why Bressius's book interested Bruno. It would appear most contradictory to find Bruno admiring the work of Bressius if it was actually, as has usually been assumed, a work on trigonometry. But this is not actually the case. The particular characteristic of Bressius's book, and one that clearly appealed to Bruno, was that it constituted an attempt to formulate a mathematics based on the number 60: *delecto sexagenario numero, vt commodissimo, quod varias partitiones absolutas admittat*.[38] At the end of his work, Bressius devises a curious and complicated table called a *Canon Hexacontadon* for facilitating calculations based on an arithmetic that stops at the number 60. Bruno refers to this as "the oracle of Bressius" so it seems unlikely that he is taking up a critical attitude to this book.

At this point, two questions need to be asked: first, "What's so special about Sixty?"[39] Second, what was it about a mathematics with base 60 that appealed to Bruno? The answer to the first question seems to be partly historical and partly astronomical. It goes back to the Sumerian civilization, which originated around 3200–3100 B.C. and used a counting system with steps at 5, 10, 20, and up to 50 but stopped at 59. The number 60 became a new unit known as *Gesh*, sometimes called *Ash* or *Dish*. It was considered a special number and as such was taken over by the Babylonians and the Assyrians. It is probable that astronomical motives decided the choice of the special number: the year made up of 360 days; twelve lunar months each of thirty days' duration. Furthermore, the importance of 60 is increased by the fact that it represents the interval which the sun spends in each of the constellations of the zodiac. It seems likely that a circle was normally divided into six parts and that the day

37. Aquilecchia 1993a:313.

38. Bressii 1581:1.

39. My remarks in the following paragraph are taken from Barrow 1992:64–67, which, however, contains no mention of Bressius.

was divided into twenty-four hours, the hours into sixty minutes, and so on because of the perceived convenience of the number 60. The Greek astronomers also worked in sexagesimals and are presumed to have taken over the tradition from the Babylonians, although they formulated their mathematics in terms of the decade and not, like the ancient Sumerians, on a system of 60. Bressius proposes to develop a mathematical doctrine in which, once again, 60 becomes the basic number of mathematics as well as of practical life and astronomy. He takes considerable pains to carry out this task; and although he contributed tables of sines and tangents, his work cannot be considered fundamentally a book of trigonometry. As far as I know, his book met with complete oblivion from his day to this.

Why, then, was Bruno so interested in it? The answer is clearly indicated by Bruno himself, whose numerology in *De triplici minimo* actually carries into effect a system based on sixes and their multiples. For example, at the end of book III, he proposes what he calls a "ladder of succession" proving that the division by squaring of the "natural" division of the circle into six parts will always give a finite number because the part of the curve will always maintain a finite dimension. Bruno begins his ladder with the number 60 considered as the minimum part of the circle. The meaning of this becomes clearer later on, in chapter 10 of section II of book V, where he reproduces the triangle of Leucippus indicating the mode of formation of atomic minimi into the form of a triangle. Because Bruno's atoms are all spherical, their centers are joined by straight lines that divide each circle "naturally" into six parts, or angles of sixty degrees. This basic division is maintained in what Bruno clearly thinks of as the primary atomistic formation: that of a circular body, which he calls the "Field of Democritus" and illustrates in book I, chapter 14.

Bruno here is translating into atomic terms that particular use of the number 60 which had, from prehistoric times, been noted in the domain of astronomy. He could thus restate in mathematical terms the equivalence between the maximum universe and the minimum atom on which his cosmology was founded. This sexagesimal mathematics, revived by Bressius, had solid historical foundations and appeared to be justified by the commonest rhythms and divisions of natural time and revolution. If the vicissitudes of the infinite cosmos were in some way determined by the number 60 and its multiples, so, for Bruno, were the agglomerations of the minimum atom. The number 60 is the minimum quantity of the "ladder of succession" that closes the third book of Bruno's *De triplici minimo*.

11. Diagram in Giordano Bruno, *De triplici minimo et mensura* (1591), book 1, chapter 14. Reproduced by permission of the Biblioteca Nazionale Centrale, Rome.

The other mathematical work on which Bruno based his mathematics is the *Elements* of Euclid. The way he makes use of this book is not, however, conventional and requires some explanation. First, in Bruno's book IV of *De triplici minimo*, "On the Principles of Measure and Figure," the basic elements of Euclidean geometry, the circle and the triangle, are presented through the three temples of Apollo, Minerva, and Venus, in semi-mythological terms. Only when the "meanings" of the three temples have been explicated is Euclidean geometry illustrated as the basis of all measurement in the final, and very seldom commented on, book of Bruno's mathematical work called "The Book on Measure."

It is important to understand the fourth book of *De triplici minimo* as a discourse on the "principles" of all measurement and form. The frequent use of mythology here, beginning with the names of the three temples, can easily lead to expectations of some magical or mystery treatment of mathematical entities, an interpretation that can be allowed only in the terms and to the extent authorized by Bruno himself. In the introductory page of this book he states the classical Greek and Latin dimension of his discourse. Measure, carried out with the aid of number, is the imposition of reason or "determination" on to the phenomenological world. As his entire work is designed to show, such determination finds its point of departure, as well as its founding justification, in the existence of a minimum quantity or monad. In the celebration of the monad Bruno recognizes the necessity to "visit the recesses of the sacred fountain": here lie the mysterious origins of all things, the active power of all number. In this sense, the foundations of the temples of Apollo, Minerva, and Venus can be associated with the primordial figure of the circle and related to similar figures present in ancient mystery cults. Yet Bruno is very cautious in expressing such relations in terms of religious doctrine. Rather, he defines his temples as objects which are internal to the mind. The determination and order of measurement proceeds through the ordered structure of the mind. Once the temples have been constructed on the basis of circles, triangles, and squares, the archetypal "figures" of geometry, they can be interpreted only in terms of their own definitions, which are those of Euclidean geometry. In chapter 7, Bruno outlines briefly the fundamental postulates of that geometry. Then, in one of those apparent fits of unreason which so irritated Tocco and has continued to irritate many others after him, he goes on to offer "demonstrations" of those definitions, not in terms of the conventional theorems but of classical mythology.

Let us take as an example the demonstration defined by Gorgonio: "The way by which the Graces descend towards the fields of Jupiter are suspended above the paths of Ocean and fierce Mars in such a way that the parts which are equally distant from the threshold of Jupiter remain equidistant from it, on one side and the other, forming equal angles, and the path does not incline any more towards one side than it does towards the other." This is Bruno's way of defining an isosceles triangle (see Euclid, *Elements*, book I, Theorem 2, Proposition 5), and perhaps not as "fantastic" as it has usually seemed to his critics. For this proof had already been much discussed, owing to the complicated argument produced by Euclid himself, who extended the original triangle and formed two larger ones to demonstrate that the two base angles were equal. Proclus showed that the second triangle could be constructed inside the original one, and not by extension, thus simplifying the proof. Then finally Pappus proposed another demonstration needing no addition at all because the one triangle is considered as if it had doubled itself, becoming two equal triangles. Bruno, who may have been better read in the mathematics of his time than is usually supposed, appears to be giving a mythological description of Pappus's setup, and then simply stating the theorem as true.[40]

What, one may nevertheless ask, is it all about? In my opinion, these pages must be understood both in terms of the reference to classical Greek and Latin antiquity with which this book opens and to the philosophy of mind which Bruno has just delineated as the proper context in which to understand the "temples" of geometry. He is emphasizing that the classical tradition, on which renaissance culture was firmly founded, could be defined in terms of the passion for order: an ordering of experience which is as much expressed by the Pantheon of mythological gods and stories as by Euclidean geometry. They can best be seen as two separate moments of experience. Geometry supplies the ordered mental forms of measurement which are to be projected onto the shadowy phenomenological shapes or "bodies" perceived in nature: shapes that may, as things in themselves, be no more objective or "true" than the myths of ancient classicism. That is, a problem (perhaps an insoluble one) may arise because the phenomenological world perceived by the senses, just as much as the ordering passion of mathematics, derives from categories founded in the structure of the mind. Bruno is thus making a historical and philosophical statement, as well as offering an early definition of the problem of self-referentiality which is at the center of discussion today.[41] He is also indicating the way in which science must proceed: not by identifying forms (geometry) with things in themselves, but by understanding them as systems of relationships determined by the mind. Bruno, however, was no pure Platonist; and he is not concerned with mathematical entities as purely rational phenomena existing only in a sphere of intellectual truth. He is concerned with measuring the bodies (mythology) perceived in the physical world, however desperate and vain an attempt that may turn out to be.

Book V, or "The Book on Measure," finds its tension precisely in this concept of measurement, referred in particular to the minimum particle or atom underlying the bodies perceived by the senses. Bruno is address-

40. The English translation of Euclid's *Elements of Geometrie* by Henry Billingsley, published in London in 1570, gives all three of these proofs of this theorem. It is of some interest to know that a computer which was programmed with all of Euclid's definitions and axioms, and then asked to come up with the proofs, produced the Pappus proof of the isosceles triangle and not the more complicated proofs of Euclid himself or Proclus. See Boden 1990:75–111.

41. See Hofstadter 1985:5–48.

ing a question similar to those being raised by modern quantum physicists and mathematicians. This is not to be unhistorical. Bruno was working at the very beginning of the modern revival of ancient atomism and was not content with merely formulating an atomistic theory of matter. He goes further, posing the fundamental question, Is the minimum particle of matter subject to normal physical laws of order and measurement? Containing within itself, in a possibly incalculably compressed form, the energies of the universe, is it going to be measurable at all in terms of the logical capacities of the human mind? The power and insight which lie behind this question, asked in 1591, can be judged by the fact that the answer is by no means clear even today.[42]

My discussion of Bruno's mathematics has centered on those works and passages where he considers, often with perplexity and at times with desperation, the circular process set up by the human mind in its scientific endeavor: the definitions and axioms it defines in order to understand and measure bodies in space, and the consequent attempt to apply those definitions to a phenomenological world itself closely linked to, and possibly dependent on, categories of human understanding. Bruno's realization that the problem could only be exacerbated by his own cosmological and atomistic theories, which call into account minimum and maximum entities tending, by definition, to elude the finite perceptual and intellectual capacities of the human mind, is what makes his mathematical uncertainties, in my opinion, so interesting and relevant today.

In his essay "Rediscovering the Mind," Harold J. Morowitz has written of the creation of twentieth-century physics:

> With the development of quantum mechanics, the role of the observer became an even more central part of physical theory, an essential component in defining an event. The mind of the observer emerged as a necessary element in the structure of the theory. The implications of the developing paradigm greatly surprised early quantum physicists and led them to study epistemology and the philosophy of science. Never before in scientific history, to my knowledge, had all of the leading contributors produced books and papers expounding the philosophical and humanistic meaning of their results.
>
> Werner Heisenberg, one of the founders of the new physics, became deeply involved in the issues of philosophy and humanism. In *Philosophical Problems of Quantum Physics*, he wrote of physicists having to renounce thoughts of an objective time scale common to all observers, and of

42. For the present crisis in the field of quantum measurement, see Penrose 1989.

> events in time and space that are independent of our ability to observe them. Heisenberg stressed that the laws of nature no longer dealt with elementary particles, but with our knowledge of these particles—*that is, with the contents of our minds.*[43]

But what of Bruno's contemporaries? Writing as he did in the early modern years of post-Copernican and preatomic speculation, was he not ignoring the primary necessity of the time: that of measuring and reducing to order our own limited corner of the new, unbounded universe? It is surely their willingness to dedicate themselves with patience and obstinate humbleness to precisely this task which distinguishes Galileo, Tycho Brahe, Kepler, and later on Boyle and even Newton, who have left only in private papers and notes the larger epistemological and metaphysical questions and doubts that at times assailed them. Bruno's willingness to face those questions, at such an early stage and with such speculative daring, gives his mathematical pages their particular flavor and interest. When he attempted to follow a more orthodox and strictly technical procedure, he was like a bird with clipped wings, as the two recently discovered mathematical texts, hurriedly put together in his last days of freedom, sadly suggest.

In 1960, Paul Oscar Kristeller communicated the discovery of two unknown Brunian manuscripts held in the Universitätsbibliothek of Jena. The question was taken up by Giovanni Aquilecchia, who recognized in those manuscripts two so far unknown and unpublished works on mathematics titled respectively *Praelectiones geometricae* and *Ars deformationum*, published by Aquilecchia in 1964. In his reconstruction of the history of these two brief Latin works, Aquilecchia established their composition during the few weeks Bruno spent in Padua soon after his return to Italy in the autumn of 1591.[44] Although documentary evidence is almost nonexistent, it appears certain that Bruno was privately teaching a group of German students in Padua closely connected to J. Besler, who had already established himself as an admirer of Bruno during his stay in Helmsted. In Padua, Besler is known to have copied out for Bruno several manuscript works, now in the Noroff collection in Moscow, which were published by Tocco in the third volume of the national edition of the Latin works in 1891. These include the final works on magic as well as the long and important *Lampas triginta statuarum*, composed at Witten-

43. See Morowitz 1981:34–49.
44. See Aquilecchia 1993a:213–36.

berg in 1587, which has been at the center of much discussion of Bruno's thought.[45] Aquilecchia convincingly associates the newly found mathematical texts with the copying being done by Besler and his fellow students at Padua, further adducing the possibility that the episode should be seen in the light of an attempt by Bruno to be appointed to the chair of mathematics left vacant by Giuseppe Moletti in 1588 and awarded to Galileo in September 1592. Aquilecchia restated this thesis with further arguments in his more recent work on Bruno's mathematics; and although no known documents exist to show that Bruno ever approached the academic authorities in this sense, it does seem to fit in with the characteristics of the new mathematical texts.

Both the *Praelectiones geometricae* and the much briefer *Ars deformationum* are reelaborations of parts of book IV of *De triplici minimo*. In fact, the *Ars deformationum*, to be understood as the art of form and not of deformation, is a pamphlet of instructions showing how to construct the complicated geometrical figures that constitute the temples or seals of Apollo, Minerva, and Venus. Although collected in the Jena manuscript after the *Praelectiones*, it should presumably precede it, as the seals are proposed by Bruno in *De triplici minimo*, book IV, chapters 3, 4, and 5, whereas the *Praelectiones* discusses definitions elaborated in book IV, chapters 8 and 9. The *Praelectiones* is characterized above all by an apparent desire for academic orthodoxy according to the principles of the time. If it contains no new material with respect to *De triplici minimo*, it presents the two chosen chapters in terms that severely reduce the mythological dimension of the earlier text, while at the same time eliminating, apart from one brief reminder of the triangles of Leucippus and Democritus, any reference to the atomic minimum. This double reduction serves to indicate the way in which myth and atomism seem to have been connected in Bruno's mind: the two schematic representations of an essential, unseen order inherited by the modern world from Greek philosophy and civilization. Once the unorthodox cosmology and the dangerous atomistic theory of matter have been eliminated, presumably in the light of the intellectual realities of the academic life of the time (it should be remembered that Galileo taught astronomy in Padua in the 1590s in strictly geocentric terms), one may legitimately wonder what is left.

In chapter 8 of book IV of *De triplici minimo*, Bruno had very rapidly proposed the fundamental principles determining the unity, identity, similitude, and proportionality of the basic figures of geometry. In the

45. See in particular Ciliberto 1990:205–11.

prose comment on his definitions, found in the poetic pages of his text, he had indicated the precise parts of Euclid's *Elements* that he had taken them from. In chapter 9 he had referred to some of the principal Euclidean theorems, again citing in the prose comment the precise passages from the *Elements* that he was using. These chapters indicate that measurements in space require a geometry as a conceptual apparatus to which they must adhere; and the available geometry to which Bruno refers is Euclidean. The real sense of his discourse emerges, however, in book V, where the act of measurement in terms of Euclidean geometry is considered with respect to atomic minimi and their termini. Here fundamental questions concerning the measurement of minimum particles are asked, which are still relevant today: if the minimum particles are alike, and all are imbued with powerful forms of energy, how will it be possible to keep track of one among many?[46] Will it be possible, in the act of measurement, to distinguish the limits or termini from the atomic minimums themselves? But such questions could hardly be asked in 1591 in an academic course where neither the infinite universe nor the atomic minimum could be mentioned. Deprived of the most original outcome of his geometrical doctrine, all Bruno could do was to enunciate the axioms of Euclid as clearly as possible. He introduces them with an entirely orthodox page of Aristotelian doctrine, based on the *Posterior Analytics*, book II, where the various modes of inquiry are clearly distinguished into three types: demonstration by cause (which belongs to physics), demonstration *simpliciter*, that is, by definition (which belongs to mathematics), and demonstration by signs (which belongs to metaphysics).[47] Bruno is concerned here with demonstration *simpliciter*; and what remains is a very elementary textbook of Euclidean geometry, not particularly enhanced by its failure to mention Euclid. This was all that Bruno had to offer to the still narrowly neo-Aristotelian academic culture of his time.

46. In modern particle physics, many types of minimum particles are now recognized: neutrons, positrons, quarks, and so on. All these particles have different characteristics, but within each class all are equal and identical to the others. This is now recognized as one of the major problems in the acts of observation and measurement of the particles themselves. See Penrose 1989:291–390.

47. Aristotle originally divided his explanation into four types, but the later tradition of Aristotelian commentary had already reduced these to three: see Aquilecchia's introduction and his note on this page of the *Praelectiones geometricae* in Bruno 1964:xviii–xxiii and 3.

10 Epistemology II: Picture Logic

In the *De triplici minimo* Bruno investigates the possibility that numbers and Euclidean geometry can act as the mnemonics required to reveal the nature of finite bodies composed of atomic minimums in a universe of infinite space. Then, at the very end of the book, he seems to lose faith in number as the tool for determining quantity, to a large extent, according to Hélène Vedrine, because his number theory was limited to the Pythagorean decade of natural numbers and also because he rejected trigonometry: both factors being linked to what Vedrine calls the "realist obstacle" to Bruno's mathematics.[1] Deprived of some of the most powerful mathematical tools available in his time, Bruno would have had difficulty in measuring even the observable bodies of the natural universe. Establishing a method of defining and investigating the atomic particle was another insoluble problem. That Bruno recognized the nature and extent of the mathematical dilemmas caused by the renewal of an atomic theory of matter was undoubtedly a remarkable achievement. Not surprisingly, however, his major mathematical work leads to a dead end, and Bruno finishes the *De triplici minimo* by considering the possibility of forming symbolic reference systems not out of numbers but of letters.

Bruno was not thinking of an algebra, in which letters stand for numbers, but rather of ordered tables or patterns in which the letters of the alphabet stand in their own right as cyphers or signs in the creation of a new symbolic language comparable to that of mathematics. Although the Lul-

1. See Vedrine 1967:239–48.

lian source for this aspect of Bruno's work has long been recognized, it may have been reinforced by the *Divina proportione* published in Venice by Luca Paccioli in 1509, where, in a section at the end of the text, large capital letters of the alphabet are considered in terms of their geometrical shapes and symbolic potential. Paccioli's book, which is largely a mathematical treatment of perspective for the use of architects, constitutes an interesting possible source for Bruno's mathematical doctrine.[2] After the text, there appears a page-size illustration of a man's head with a network of geometrical figures, primarily triangles and squares, superimposed on it. There follows a series of pages, each dedicated to the letters of the alphabet; curiously it lacks the Z, which Bruno uses to terminate his own book of mathematics.[3] In Paccioli's volume, the letters are followed by one or two pages with illustrations of classical columns, followed by a page illustrating the large door of a classical temple. This page leads into the final part of the book, dedicated to carefully constructed illustrations of geometrical solids.

The mathematical constructs in Paccioli's book bear a remarkable resemblance to those of Bruno. Both the idea that the principal geometrical figures constitute a grid through which the mind approaches reality and that the temple door of mathematics is the passage through which the mind must pass in its investigation of reality came ultimately from ancient, pre-Greek sources. Paccioli had given them eloquent and strong expression, mixing the letters of the alphabet with geometrical figures as if both could be manipulated indifferently as the bases of symbolic languages and signs. Bruno took over this idea as part of his infinite and atomically structured universe, applying it to the possibility of the alphabet creating infinite combinations of messages coded to some natural grammar, corresponding to the order of the natural world. The program remains nebulous and appears only in five tables that abruptly close *De triplici minimo*. Bruno calls the tables "Mothers" and "Daughters" because one table is supposed to generate another; the last is based on the power of the letter Z, representing the boundaries of things, to cut across the whole picture of space.[4]

This, of course, is only another way of stating the atomic conception of matter. It is interesting because of the change from numbers to letters

2. I am grateful to J. V. Field for suggesting to me Bruno's possible use of Paccioli's book.

3. The British Library copy of Paccioli's text (shelf mark 50.e.20) carries, in the place where the letter Z should have been, an inserted note which reads: "There is no letter Z (zeta)—and I never saw this letter in any copy." The note is signed F. F. Molini.

4. See Bruno 1879–91, I, iii:356.

as possible tools for defining and comprehending the atomic world. As a powerful writer and poet, Bruno had always used letters and words as his linguistic tools. He had also, from his very first work, *De umbris idearum*, been investigating the possible uses of the classical art of memory, related to the Art of Raymond Lull, an inquiry which Bruno developed in later works.[5] Although all serious critics of Bruno's philosophy have dedicated time and attention to his art of memory and to its place in the development of his thought as a whole, they have not often related it to his mathematics.[6] This seems to me a mistake because the art of memory and mathematics appear to vie throughout Bruno's work as the most appropriate logical tools with which to investigate the newly defined universe. Bruno's memory images, as well as his letter charts at the end of *De triplici minimo*, should be related to the temples of Apollo, Minerva, and Venus, their pattern formed by letters or pictorial images instead of geometrical figures. Like the temples, they constitute the "doors" through which one may understand the formal structure underlying the phenomenological world.

Bruno's works on memory which I shall consider in most detail are the first and the last: the *De umbris idearum*, published in Paris in 1582, and the *De imaginum, idearum et compositione*, published in Frankfurt in the autumn of 1591, together with the last two volumes of the Frankfurt trilogy. Bruno is known to have published two works before he left Italy in 1578, both of them, to judge from the titles, written in Italian: an *Arca di Noè*, which he claims to have presented and dedicated to Pope Pius V, probably sometime before 1572, and a work titled *De' Segni de' Tempi*, which at his trial he said he had published in Venice, where he stopped on his way north in 1577.[7] Neither of these works has survived, and the various suppositions about their contents are pure speculation. One of them, at least, may have been a work on memory, as Noah's ark could serve as

5. There is no full-length work on Bruno's relationship to the Art of Lull. For Lull, see Yates 1982:3–125 and Yates 1966.

6. For both aspects of Bruno's thought with relation to his science, see in particular Bernart 1986 and 1994, to whose work I am indebted in parts of this chapter.

7. Bruno refers to other writings in his early works *De umbris idearum* and the comedy *Candelaio*: for example, three Italian titles which he gives as *Gli pensieri gai, Il tronco d'acqua viva*, and *Purgatorio de l'inferno*, and a Latin *Clavis magna*, which seems to have been another work on memory, and which he refers to repeatedly in the *De umbris idearum*. It is not clear whether these works were ever published, although Bruno speaks of the *Clavis magna* in particular as if it were a published text. The *De' segni de' tempi* was probably a pamphlet on weather forecasting: see the section on this work by Germana Ernst in Canone, ed., 1992:79–85.

Actus	Aeiouhxkqjlnrsgzcdtfwmb p
Beatitudo	Eiouhxkqjlnrsgzcdtfwmbp A
Certitudo	IouhxkqjlnrsgzcdtfwmbpA e
Consensus	OuhxkqjlnrsgzcdtfwmbpAe i
Idea	VhxkqjlnrsgzcdtfwmbpAei o
Fruitio	HxkqjlnrsgzcdtfwmbpAeio u
Habitus	XkqjlnrsgzcdtfwmbpAeiou b
Motus	KqjlnrsgzcdtfwmbpAeioub x
Lex	QjlnrsgzcdtfwmbpAeioubx k
Productio	J lnrsgzcdtfwmbpAeioubxk y q
Verbum	LnrsgzcdtfwmbpAeioubxkq j
Seruitus	NrsgzcdtfwmbpAeioubxkqj l
Auersio	RsgzcdtfwmbpAeioubxkqjL n
Acceptatio	SgzcdtfwmbpAeioubxkqjLn r
Continuitas	GzcdtfwmbpAeioubxkqjLnr s
Vsus	ZcdtfwmbpAeioubxkqjLnrs g
Miraculum	CdtfwmbpAeioubxkqjLnrsg z
Possibilitas	DtfwmbpAeioubxkqjLnrsgz c
Pœnitentia	TfwmbpAeioubxkqjLnrsgzc d
Repressio	FwmbpAeioubxkqjLnrsgzcd t
Ordo	wmbpAeioubxkqjLnrsgzcdt f
Obiectum	MbpAeioubxkqjLnrsgzcdtf w
Accessio	BpAeioubxkqjLnrsgzcdtfw m
	y
Recessus	P AeioubxkqjLnrsgzcdtfwmb

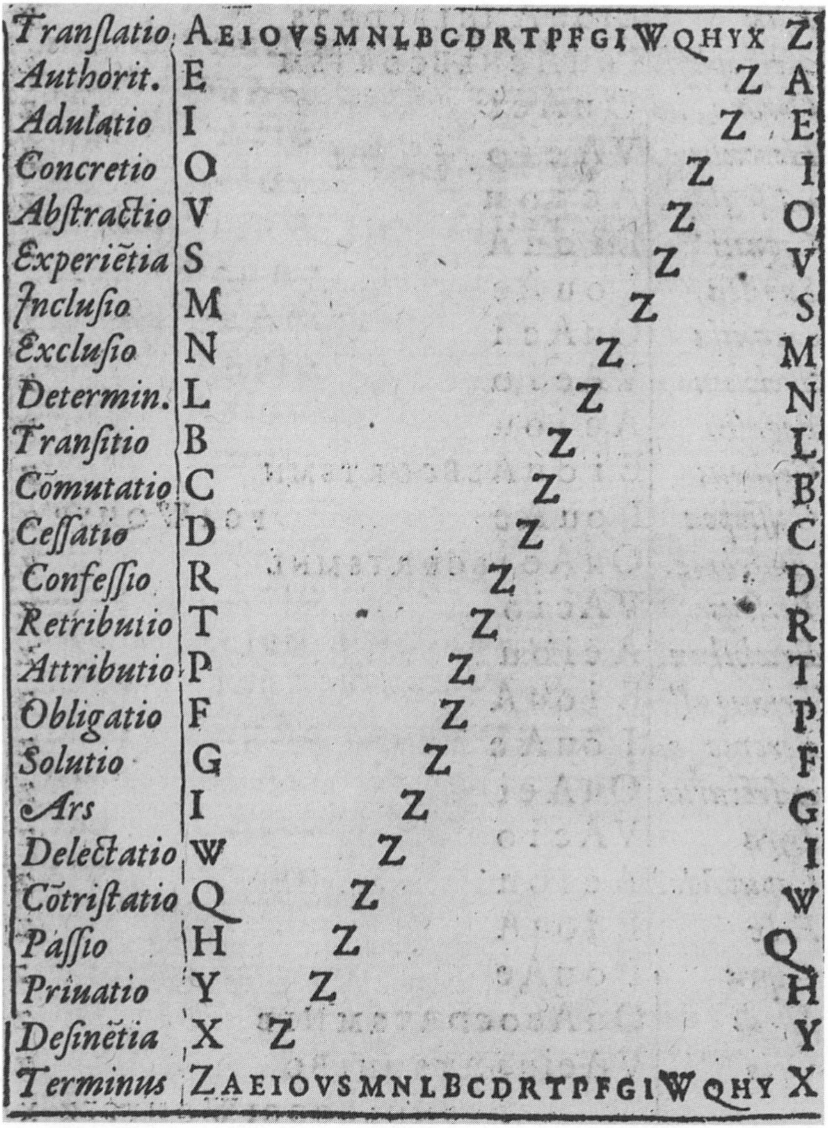

Translatio	A E I O V S M N L B C D R T P F G I W Q H Y X	Z
Authorit.	E Z	A
Adulatio	I Z	E
Concretio	O Z	I
Abstractio	V Z	O
Experiẽtia	S Z	V
Inclusio	M Z	S
Exclusio	N Z	M
Determin.	L Z	N
Transitio	B Z	L
Cõmutatio	C Z	B
Cessatio	D Z	C
Confessio	R Z	D
Retributio	T Z	R
Attributio	P Z	T
Obligatio	F Z	P
Solutio	G Z	F
Ars	I Z	G
Delectatio	W Z	I
Cõtristatio	Q Z	W
Passio	H Z	Q
Priuatio	Y Z	H
Desinẽtia	X Z	Y
Terminus	Z A E I O V S M N L B C D R T P F G I W Q H Y	X

12. Diagrams in Giordano Bruno, *De triplici minimo et mensura* (1591), book 5. Reproduced by permission of the Biblioteca Nazionale Centrale, Rome.

well as the memory theater of Giulio Camillo Delminio as a mentally conceived architectonic structure for "storing" conceptually all the forms and grades of being.[8] In 1675 the Jesuit Athanasius Kircher published in Amsterdam an *Arca Noe*, one of the most important of the seventeenth-century works on memory.[9] The actual contents of these lost works, however, is unknown. More important still is the disappearance of a work titled *Clavis magna*, which undoubtedly contained the technical details of Bruno's art of memory. He frequently refers to it in the *De umbris idearum*; and the reading of that work is made particularly arduous because no text of the *Clavis magna* remains to elucidate the far from simple art of memory which Bruno is attempting to explain.

Both his first surviving published work, the *De umbris idearum*, and his last work, *De imaginum*, published before his imprisonment and with his consent, are on the art of memory. Bruno also develops his memory techniques in *Cantus circaeus*, published in Paris in the same year as the *De umbris*, 1582, and in *Explicatio triginta sigillorum* and *Sigillus sigillorum*, both published soon after Bruno's arrival in London in 1583.[10] We can thus see that he was most concerned with the art of memory in the early period, before the publication in 1584 of *The Ash Wednesday Supper* and the other Italian dialogues in which he developed the new post-Copernican, infinite cosmology. Although the art of memory reappeared in the *Lampas triginta statuarum* and the two Lullian works of 1587–88, it seems to have lost in primary importance. It was only after the final Frankfurt trilogy that Bruno produced, almost as a last-minute surprise, his major work on memory, *De imaginum, signorum et idearum compositione*.

A question which needs to be addressed in my opinion is why we find the intense early speculation on the art of memory apparently waning once the new cosmology has been fully developed. A closely related question is why, after he had published both the major cycles of philosophical inquiry, the six Italian dialogues of 1584–85 and the Frankfurt trilogy of 1591, Bruno produced a further book on memory that is commonly considered one of his most creative works. I suggest that in those intervening

8. For another possible interpretation of Bruno's lost *Arca di Noè*, see Badaloni 1997:44–45.

9. For Kircher's work, see in particular Perugini 1984:115–20.

10. Some would also include among the works on memory the more specifically Lullian texts such as *De compendiosa Architectura et Complemento Artis Lullii* (1583), *De lampade combinatoria* (1587), and *De specierum scrutinio* (1588), as well as the *Lampas triginta statuarum*, first published from manuscript in 1891 and probably written in Wittenburg in 1587.

years Bruno was intent primarily on gauging the extent to which mathematics could or should be developed as the logical tool for an inquiry into the newly infinite and atomic cosmos. When that path seemed to be leading him nowhere, Bruno returned to the traditional art of memory as possibly, after all, the most appropriate tool for investigating the shape and nature of the new universe.

It is probable that Bruno was a Copernican before he came to England in 1583. His Parisian works mention the sun as center of the universe, although they contain no technical developments of the Copernican proposal.[11] These developments follow his arrival on the banks of the Thames, with the accompanying illumination resulting from the opening of the vase of Circe by the English nymph. The illumination, it will be remembered, was of a double sort, symbolized by the two eyes of the nymph. The celebration of the newly perceived natural universe which closes the *Heroici furori* is thus of great interest and importance. On one side, the universe is presented as ordered, not random in its processes: an order expressed by the harmonious music and song of the newly seeing seekers. On the other, the double power of the nymph's eyes suggests the possibility of distancing themselves from the natural processes and comprehending their pattern with the mind.[12] The immediate problem, then, is how, in the vastly expanded spaces of the new infinite universe, such a process of comprehension is to take place. Barrow presents a modern formulation of the same problem:

> If the sequence of experiences is not random, then it will possess some pattern which permits the information within it to be represented in an abbreviated form without any degradation of its content. If no such mnemonic is available, the sequence of experiences would be regarded as random and it could not be communicated in any way except by a complete repetition. Science is the search for these abbreviations, or "compressions", of the fruits of experience. This quest is remarkably successful, a fact that bears witness to the extraordinary compressibility of the Universe in which we live. Indeed, perhaps we could only survive in a universe that possessed a fair measure of compressibility which our brains could take advantage of to store information about the environment for future use. We can see why any such compression of experience relies upon those things that we call "symbols". Compressions seek to store lots of pieces of information in a single mark or representation. Symbols are the result of such a desire and mathematical

11. For the history and significance of the renaissance cult of the sun, see Garin 1975.
12. See Bruno 1958:1165–78.

> symbols have proved to be the simplest and most succinct way of storing information.[13]

The final part of this statement would surely have been approved by all those about to develop the new classical science, from Kepler to Galileo, from Descartes to Newton. It seems that Bruno, although attempting to do so in a rather personal way in *De triplici minimo*, was unable in the end to endorse it. Was Bruno's failure with mathematics entirely owing to his rejection of some very basic techniques which he seems not to have understood, or did that rejection itself arise from some deeper motive of distrust such as we find, for example, in Robert Fludd? Did his return to the traditional art of memory in *De imaginum, signorum et idearum compositione* succeed in presenting it as a logic that has any relevance for the scientific endeavor, or should his art of memory be seen throughout as relevant only to other interests, whether magic and the occult or oratory and art?

In *The Art of Memory*, arguably her finest work, Frances Yates elaborated a remarkable historical overview of the traditional art of memory from classical times until its waning in the seventeenth century. She claimed that what in classical times had been initiated as a mnemonic technique in the context of public life and oratory became in the renaissance a process for ordering, in terms of imaginary architectonic memory "places" or "theaters," data and knowledge of the cosmos. Bruno's works on memory, considered by Yates as one of the highest points in this development, were thus discussed in a context that might seem to lead toward his science and new cosmology.[14] Yates, however, never fully developed this line of thought. Her book on the Hermetic tradition had already excluded Bruno from the history of renaissance science, transforming all his thought on the natural universe into magical doctrine in the occult tradition of Florentine neo-Platonism. What interested Yates in the *De umbris idearum*, for example, was Bruno's use of images of the signs of the zodiac and his Lullian memory wheels composed of numbers and letters from the ancient alphabets. She believed such images and icons were constructed to contain magic energies and powers that could be manipulated to call down into the mind the higher grades of being and knowledge contained in the stars. Through the influence of these "superior agents," the Magus could learn about the nature of the things in the lower world or earth. The difficulty is that Bruno, from the beginning, refused to con-

13. See Barrow 1992:23–24.
14. See Yates 1966.

template the neo-Platonic concept of hierarchical grades of being in the natural universe on which such an interpretation of his art of memory depends. Although I recognize the historical merits of Yates's book and the force of an interpretation that undoubtedly illuminates many other examples of renaissance memory techniques, I am convinced that she failed to understand the extent to which Bruno himself had struck out along very different paths. The question remains: What use did he contemplate for the classical and renaissance art of memory within the newly infinite spaces of the post-Copernican universe he was gradually coming to understand?

Rita Sturlese has clarified an essential point unobserved by Yates: that the constructions of memory places in both the *De umbris idearum* and the *De imaginum, signorum et compositione* are designed in very complex ways so that they function similarly to calculatory tables: that is, they can be used for the formation of words, or even phrases, linked to images designed to help memorize them.[15] This discovery throws new light on the question. To begin with, the series of images taken from the stars or the halls of the zodiac, or even the letters and figures of the various alphabets, are clearly not functioning with their usual meanings, either explicit or occult. They are functioning as ordered series of cyphers in much the same way that letters may be used to clarify a geometrical figure without retaining any of their meaning as letters used to form words. Sturlese admits that to obtain this surprising result in her reading of these texts, she altered some of the cyphers in Bruno's works which the printer seems to have mistaken. No doubt Bruno taught in classes or with private groups of students, using manuscript versions of his memory tables without such mistakes. In any case, Sturlese has succeeded in reelaborating them in what must have been their original versions, and her reconstructions leave little room for doubt that Bruno's memory wheels and tables were indeed technical instruments intended for practical use.

What exactly were they used for? Sturlese has been unable to answer this question. She has insisted that her findings prove the Yatesian magical interpretation of Bruno's works on memory to have been mistaken. This conclusion is supported by Bruno's statement at his trial that when the French king summoned him in Paris to ask him if his tables of memory worked by magic, he replied that they did not: they worked by art (*scientia*).[16] What "art" was intended and for what practical purposes is not

15. See Sturlese 1990 and 1992b.

16. ". . . acquistai nome tale che il re Henrico terzo mi fece chiamare un giorno, ricercandomi se la memoria che havevo et che professavo era naturale o pur per arte magica;

clear. The lively discussion in London between Bruno's friend Alexander Dicson, who published in 1583 a *De umbris ratione*, remarkably similar to Bruno's own *De umbris idearum*, and his opponent William Perkins of Cambridge, an ardent disciple of the reformed logic of Ramus, suggests that Bruno's art of memory was being associated with his scientific speculation. Perkins's replies to Dicson in his *Antidicsonus* and *Libellus De memoria*, both published in 1584, have been amply discussed elsewhere and need not be elaborated here.[17] What does need to be underlined about this English discussion of Bruno's art of memory is the way it links it securely to his cosmological dialogues. Dicson, who is mentioned as one of the guests at the Ash Wednesday Supper, is one of the most clearly approved speakers in Bruno's second Italian cosmological dialogue *Of the Cause, Principle and One*.[18] Perkins attacks this link between the new astronomy and the art of memory, claiming that there is no relationship between an art of memory based on traditional imagery such as that of astrology and a true astronomical science.[19] For Perkins, the art of memory is a general doctrine of signs, whereas astronomy needs a special language in which to investigate the movements of the heavenly bodies. It is important in this context that in Bruno's Italian works written in London, where he develops the post-Copernican, infinite universe for the first time, a very limited mention is made of mathematics as the logical tool

al qual diedi sodisfattione; et con quello che li dissi et feci provare a lui medesmo, conobbe che non era per arte magica ma per scientia." In the second hearing of Bruno at Venice, May 30, 1592. See Firpo 1993:161–62.

17. See Yates 1966:271, Ricci 1990:19–27, and above all Aquilecchia 1991:87–97. The intimacy of Bruno's friendship with Dicson has been further underlined by the discovery, made by Rita Sturlese, of a copy of Bruno's *De umbris idearum* held in the library of University College, London, which contains a personal dedication to Dicson. See Sturlese 1987.

18. ". . . dotto, onesto, amorevole, bencreato e tanto fidele amico Alessandro Dicsono, che il Nolano ama quanto gli occhi suoi": Bruno 1958:214.

19. Vides opinor artem tuam nugatoriam et puerilem esse: ideóq; decerpis ac delibas ex omni scientiarum genere flores quosdam, quibus illa distincta esset et perspersa. Quod si dixeris ex Zodiaci signis et lunae circulis, suppeditari nobis immensam quandam umbrarum supellectilem, erroris et stultitiae insimulandus es. Nam

> *Artes consequentes non sunt antecedentibus necessariae, quia sine iis hae docere possunt non contra.*
>
> *Ars memoriae antecedit, signorum et circulorum lunae doctrina sequitur. Generales artes natura priores sunt, speciales natura posteriores.*
>
> *Ars memoriae generalis est, signorum doctrina, id est Astronomia speciales est, quia illa circa rem omnem, haec circa coelum versatur. Ergo nec Astronomia, necque illius pars ulla necessaria est ad artem memoriae.* (Cantabrigiense *Ad Alexandrum Dicsonum admonituncula de artificiosae memoriae quam publicè profitetur vanitate*, London, 1584, fol. D7v).

needed to investigate the much enlarged cosmos. Even the atomism, although already formulated, is still in an embryonic stage. All Bruno's attention is dedicated to establishing the reasons for accepting the new Copernican astronomy and above all its extension to an infinite universe inhabited by an infinite number of worlds. The problem is how this new unbounded universe is to be visualized and understood.

Dicson's active participation in Bruno's cosmological dialogues suggests that both of them considered the art of memory the proper tool with which to understand the vicissitudes of the newly infinite world. Even as late as 1588, when Bruno published in Prague his text *Articuli adversos mathematicos*, he was primarily concerned with the negative aspects of the new mathematics, and above all the new trigonometry. The attempt to align himself with a properly mathematical concept of method, even if in very personal and only partially successful terms, is limited to the later works in which the atomism is at the center of attention. It would thus make sense for Bruno to return to the art of memory immediately after the *De immenso*, the final cosmological work of his maturity. If I am right in this deduction, the question that needs to be addressed is: How effective could the traditional art of memory be in describing the newly immense cosmos?

Frances Yates wrote an eloquent description of the renaissance passion for order in the form of memory images or "theaters of the mind." In particular, she drew attention to Giulio Delmino Camillo's *Idea del theatro*, in which the mental space of an imaginary theater, which Camillo actually reconstructed, was designed to contain all that was remarkable in the universe of knowledge.[20] The idea, as Yates points out, would lead to eighteenth-century encyclopedism, although Camillo's version stressed the importance of mnemonic imagery and memory places rather than relying only on reason and words. It is not clear whether Bruno knew the work of Camillo, but he did know another work by a countryman of his who is mentioned in *The Ash Wednesday Supper* as an Italian exile in London when Bruno was there. Not a very renowned figure, even in his own time, Alessandro Citolini had published in Venice in 1561 a work titled *Tipocosmia*, which contains some lively satire of attempts such as Camillo's to collect knowledge of everything in the space of one memory "theater."[21] Citolini underlines with considerable effect the fundamental problem of works of memory of this kind, particularly now that the new universe had become so large. That is, can they be anything other than

20. See Yates 1966:135–62.

21. For Bruno's references to Citolini in the *Supper*, slightly different in the two known versions of the text, see Aquilecchia 1991:40–41.

repetitive catalogs of the collective order of things as already perceived by the mind? Let us consider Citolini's often rather frivolous book in the light of this very serious question.

It is situated "in a beautiful garden near the wonderful city of Venice," where a group of noblemen are discussing the pattern of life and the universe. One of the chief participants is called Count Collaltin di Collalto: an aristocratic form of repetition which seems to anticipate Citolini's problem and theme. They decide to divide their work into seven days: just as God took seven days to create it all, so they will take seven days to recreate the pattern of being and get it into a book. They do not seem worried about the possible heretical implications of their enterprise, but then, Venice was among the more illuminated societies of the time. They start off happily making list after list of the wonderful variety of things but soon run into problems. For example, at one point Citolini decides that all the numbers will have to be included because everything needs to be counted. So he starts getting them all down, and from 1 reaches, after quite a long time, the number 1000000000000. At that point he asks his reader to take it on trust, remarking that "by continuing in that way you can go on to infinity; but given that the present number would be enough for me if only it was in money, I will finish there."[22] Later on, a similar problem arises when they start doing geography and have to get down all the place names in the world, not to mention the universe. Here the Count heroically starts off in alphabetical order, which at least reduces things to a mental scheme. He is soon interrupted by a fellow nobleman who thinks it would be more interesting to follow the place names in their true order, as they lie on the map. The Count agrees and starts off in Piedmont. Nearly thirty pages later he has only got as far as Capodistria: thirty pages of lists of place names just for the very northernmost strip of Italy. At last his friends invite him to desist: "Deh! (said Ferrò at this point) please don't fatigue yourself further, my dear Count. That's more than enough."[23]

Citolini's book, whether serious in intention or not, anticipates a problem known today as self-referentiality, which occurs when those involved in computerization realize that their programs often constitute a so-called closed loop in which the mind is fed back with data the mind already knows. For, of course, the mind elaborated the program in the first place. Douglas R. Hofstadter, in a book with the suggestive title *Meta-*

22. See Citolini 1561:310.

23. Ibid., 350–79; my translation. For a treatment of the more serious aspects of Citolini's *Tipocosmia* see Bolzoni 1995:257–59.

magical Themes: Questing for the Essence of Mind and Pattern, offers his readers a self-referential story that reads remarkably like Citolini's book. The story is titled "This is the Title of this Story, which is also found Several Times in the Story Itself." Here is a sample:

> This sentence . . . reminds the reader that such allusions to a story that doesn't, in fact, yet exist are no substitute for the real thing and therefore will not get the author off the proverbial hook.
>
> Paragraph. Paragraph. Paragraph. Paragraph. Paragraph. Paragraph. Paragraph. Paragraph. Paragraph. Paragraph. *Paragraph*. Paragraph. Paragraph. Paragraph.
>
> The purpose. Of this paragraph. Is to apologise. For its gratuitous use. Of. Sentence fragments. Sorry.[24]

Scientists and philosophers of science are still puzzling over the realization that an increasingly sophisticated and specialized science can provide only a fragmented picture of the world and that such fragments are only too often uncannily recognizable as forms projected on to the universe by the mind in the first place.[25] The dilemma had become apparent by the first two decades of this century, following Einstein's new theories of relativity and the birth of quantum physics and was eloquently expressed by A. S. Eddington: "We have found that where science has progressed the farthest, the mind has but regained from nature that which the mind has put into nature. We have found a strange foot-print on the shores of the unknown. We have devised profound theories, one after another, to account for its origin. At last, we have succeeded in reconstructing the creature that made the foot-print. And Lo! it is our own."[26]

Let us get back to Bruno, who knew Citolini and mentions him in *The Ash Wednesday Supper* as an unfortunate Italian gentleman to whom the English had been nasty. They seem to have broken and crushed Citolini's arm; but when he appealed to the magistrate, it was deemed impossible that such a thing could have happened in such a respectable place.[27] Such at least is Bruno's story. No mention is made of Citolini's work; but it is possible that Bruno had been influenced by Citolini's defense of the Ital-

24. See Hofstader 1985:37–41.

25. Ibid., 232–59, "Variations on a Theme as the Crux of Creativity": where fresh situations get unconsciously framed in terms of familiar concepts; those familiar concepts come equipped with standard knobs to twiddle; twiddling those knobs carries you into fresh new conceptual territory.

26. Eddington 1920:201.

27. See Bruno 1958:77.

ian language, a *Lettera in difesa della lingua italiana*, published in 1540, to write himself in the vulgar rather than in Latin, while in London.[28] He would almost certainly have known the *Tipocosmia* and been aware of the dangers it represents. It may thus be no coincidence that Bruno immediately stresses the necessity for compression, announced in the image of the painter and his universal picture developed in the "Argument" of the second dialogue of *The Ash Wednesday Supper*, where he first proposes the new infinite universe. The painter can no longer contain within his canvas all the data with which nature supplies him and will have to be content with putting in "a head here, there a horn, in one place part of the hind-quarters, in another the ears, of this one only the whole description."[29] The problem then becomes what method to use to carry out this compression of the picture of the system.

The crisis caused by the new Copernican cosmology resulted in the disappearance of a traditional and deeply rooted visual image of universal unity and harmony. This image was based on Euclidean geometry, the language in which the astronomy of Ptolemy was written; but through the centuries it had become surrounded by a complex imagery from the arts of astrology and alchemy and the doctrine of magical correspondences between earthly and heavenly things. Perkins maintained, in his attack on Dicson, that a modern astronomy had nothing to do with the traditional visual imagery of, say, the signs of the zodiac, but had to rely on purely rational concepts. He had no need to support his thesis with a reference to the new heliocentric cosmology (Perkins, it should be remembered, occupied an important academic post at Cambridge which he proudly announced on his frontispiece) for his line of thought could equally well be seen as taking the Ptolemaic system back, in the wake of Ramus's conceptual logic, to the purity of its Euclidean origins. It can also be connected (although Perkins himself would not have approved of doing so) to the diagram of the new universe of Copernicus, which rigorously eliminated any pictorial reference to the sun or planets, delineating

28. For Bruno's adoption of the vulgar tongue in his Italian dialogues, see Aquilecchia 1993a:41–65.

29. Bruno 1958:16–17. Ciliberto 1990:67–70 has written on the link forged by Bruno between painting, philosophy, and cosmography, although he approaches the subject more in terms of the variety of life forms included in the picture than the need for compression, which seems to me to be the essential point made by Bruno. Of undoubted importance, instead, is Ciliberto's insistence on the primacy accorded by Bruno to the "eye," to observation of the smallest detail, in the new inquiry: "un lavoro sull'occhio, sul vedere naturale, contrapposto al 'cieco vedere' di 'filosofi volgari,' 'optici,' 'geometri,' 'matematici.'"

schematically only their circular orbits or spheres. In the following books of *De revolutionibus*, Copernicus would indicate the symbolical language in which his new universe would be represented as, once again, that of traditional Euclidean geometry: indeed, the fact that Euclidean geometry survived the cosmological revolution to become the "new language" of the post-Copernican science until its unique status as a description of physical space was questioned in the late nineteenth century is a tribute to the genius of the Greek geometers. Galileo and Descartes would firmly indicate it as the conceptual foundation on which the edifice of the new science was to be erected. Even Kepler, although more attached than his Mediterranean counterparts to a neo-Platonic cosmological structure of hierarchically arranged being, nevertheless identified the archetypal images not in the many cosmological "pictures" of a Robert Fludd but in "mathematical diagrams keyed with letters," or geometry, which "is a kind of archetype of the world."[30]

This dilemma of a rapidly evolving scientific imagery created an acute problem for Bruno. First, he realized that the ultimate elements of being as he conceived of it, that is, the infinitely large cosmos on one hand and the infinitely small atom on the other, were by definition nonvisualizable in conventional terms. Second, from the first stages of his reading of Copernicus, he had felt uneasy with what he thought of as the overconceptualized and excessively abstract description of the new universe couched, in *De revolutionibus*, in purely mathematical terms. These anxieties lie behind Bruno's initial declaration in *The Ash Wednesday Supper* that the foundations of his science are Pythagorean, particularly in view of the fact that his reading of Pythagoreanism, as we have seen, has to be distinguished from the neo-Platonic interpretations of a Iamblichus or a Porphyry. For Bruno the Pythagorean numbers were ways of seeing quantity not à la Cusanus or à la Kepler as superior kinds of *entia rationis*, or even as "archetypes of the world," but rather as building blocks deeply embedded in the essence of being as it is perceived by the mind. And he seems to have felt that those primary modes of perception must be elaborated into new symbolical languages closely linked to visual imagery and thus to sensible experience, understandable to the collective consciousness.

As we have already seen, Bruno's infinite cosmology denies the traditional distinction between *potentia absoluta* and *potentia ordinata* in

30. Kepler's polemic with Fludd on the question of cosmological imagery has attracted much attention. See in particular Copenhaver 1990:282–86.

God: "Now contemplate the first and best principle which is all it is possible to be, and would not be all if it was not possible for it to be all: in it, then, act and possibility are the same thing. But this is not true of other things which, although they are what it is possible for them to be, could perhaps also not be so, and certainly could be other, or otherwise, than they are."[31]

On the same page Bruno refers to the infinite universe of "other things" as defined by a certain disposition of the things themselves (in space) and a certain succession one after the other (in time). The problem is to comprehend the order of the things in space and time, a question on which Bruno feels the need to differentiate himself from both Plato and Aristotle. In the *De umbris idearum*, the first of his works on memory, he writes:

> Plato did not contemplate the ideas of single things, but only of species of things. That was both because the ideas relate only to the production of forms, and not to matter, and also because cognitive intention is concerned only with forms and not with the genus or individuals. The Theologians propose ideas of single things, because they assert that God is the cause of all, both what is proper to form and what is proper to matter. As far as this subject is concerned, we too desire ideas of single things because we conceive of everything in terms of its idea according to the universal appearance of that which can be figured and apprehended: whether that appearance is before the thing, or in the thing, or is the thing, or is after the thing: and also whether it appears at the level of sense-experience or at the level of intellect, either practical or speculative.[32]

The shadows of the ideas, then, are the shadows or imprints in the mind of the ideas of single things, subject to space and time.[33] On this point, Bruno is content to see eye to eye with the theologians. Indeed, he will always claim that his philosophy is the foundation of the true religion; for if it denies the incarnation of any one godhead, that is only because everything that appears to us in the process of flux is fully imbued already with the ordering principle of the divinity on which the whole depends. Evil,

31. Quoted in Granada 1994; my translation.

32. *De umbris idearum*, twenty-ninth concept: see Bruno 1991a; my translation.

33. Bruno's shadows of ideas can be understood in terms of the distinction made by Vere Chappell between ideas$_o$ and ideas$_m$, the former being ideas as objects of thought and the latter ideas as mental acts or events. See Chappell 1986: 177–98. According to this distinction, Bruno's ideas of single things would be ideas$_o$ and his shadows of ideas would be ideas$_m$.

in Bruno's philosophy, is strictly associated with the relative position of the individual mind in the process of time and space. If things were fully "seen" and understood, evil would disappear; the process of salvation from evil is therefore itself dependent on intellectual endeavor.[34]

Shadows of ideas of single things, then, are the stuff of which the new science will be made—not the universal ideas of Plato that cannot yield knowledge of the particular happenings of nature, but not, either, the universal logical entities of Aristotle. For if Bruno shared with Aristotle the refusal to accept Plato's transcendental sphere of ideas and the desire to link the gnoseological process with the vicissitudes of things, he disagreed with the conclusion of the *Posterior Analytics* that the "difference" which occurs when knowledge is "grasped in the mind" through memory of perceptions depends on a process of logical-linguistic abstraction. Aristotle's claim that, although one perceives the particular, perception, in the sense of understanding, depends on the logical universal had given rise, in Bruno's opinion, to that over-subtle analytical logic, obsessed with the order of words, which the new science (and here Bacon, Galileo, and many others of their generation were of the same opinion) would have to abandon.[35]

It is interesting to wonder here whether Bruno's doctrine of knowledge was influenced by the works of the Portuguese Franciscus Sanches, and particularly his skeptical masterpiece *Quod nihil scitur*, first published in 1581.[36] Sanches taught medicine at Toulouse throughout the years Bruno taught there and obtained a regular post in 1581, the year Bruno left for Paris. He gave Bruno a dedicated copy of his book, thanking him for his good company. Bruno reacts scathingly to the gift, however, writing two marginal notes on the frontispiece and first page which call Sanches a presumptuous ass and wonder how he can be found fit to teach. Andrzej Nowicki, who discovered this interesting volume in 1967, is convinced that Bruno failed to read beyond the first page, perhaps because he considered the book a message of total skepticism that precluded knowledge of natural things.[37] Nowicki points out that this would actually be a misreading of the text, as the "*nihil scitur*" of Sanches means "*ignoramus*" and not "*ignorabimus*." Sanches can be seen as preceding Descartes in

34. This idea will be taken up forcefully by Bacon.

35. For a modern defense of Aristotle from such accusations, based on the claim that the *Posterior Analytics* is not a methodology of research but rather concerned with the organization and presentation of the results of research, see Barnes 1975.

36. See Sanches 1581/1988.

37. See Nowicki 1967.

his desire to clear the ground of false theories and prejudices in order to begin a new form of research founded on a truly scientific method. In any case, whatever the reason for Bruno's brutal comment on receiving the gift from Sanches, the tone of the dedication suggests that there had been conversations between the two; and in spite of his irritation (possibly because Sanches was awarded a teaching post he himself desired) Bruno's doctrine of knowledge gained something from the Portuguese philosopher's convictions.

Sanches mounts a scathing attack on the Aristotelian doctrine of universals. He claims that universals do not occur in nature, except in individuals themselves. Thus the concept of the universal is false unless it both includes and affirms all the things contained in it, according to their true nature.[38] Sanches prints in capital letters on his page: "KNOWLEDGE IS PERFECT UNDERSTANDING OF A THING." He goes on to claim that knowledge consists of the thing to be known, understanding (*cognitio*), and the perfection of knowledge. The things to be known are infinite, and all are linked together in such a way that no single thing is prevented from hindering or helping another. Therefore, to understand one thing perfectly, it would be necessary to know everything, which is impossible.[39] Sanches says about the idea of infinity: "I can grasp 'infinity' without ever comprehending an end, but in the midst of my reflections about it I am forced to pause when I reflect that 'infinity' is an object that I shall never succeed in finally grasping, even if I add and add for ever, and make mental pictures, extended for ever."[40] In spite of the skepticism of these conclusions, Sanches concedes that knowledge of particular things, although it will never be perfect, can and should be sought.

This passage from universals to single things, or from metaphysics to physics, which was common to many philosophers of the sixteenth century, has been traced by historians of renaissance philosophy to the humanistic tradition stemming from Valla and his great admirer Erasmus, who encouraged a new philology based on a close, critical reading of texts, in opposition to the dogmatic abstractions of the metaphysicians.[41] Between Erasmus and Sanches, this position had been developed persuasively in Italy by Mario Nizolio in his *De veris principiis et vera ratione philosophandi contra pseudophilosophos* of 1553.[42] Nizolio develops a

38. See Sanches 1581/1988:213–15.
39. Ibid., 206.
40. Ibid., 243.
41. See Kraye 1993:16–69.
42. On Nizolio see Rossi 1953:59–92, 99–121. Also Kraye 1993:43–44.

radical form of nominalism derived from William of Ockham that derides both the transcendental ideas of Plato and the abstract universals of Aristotelian logic. Sweeping aside the Peripatetic tradition from Boethius through Albertus Magnus, Thomas Aquinas and Duns Scotus, Nizolio attacks the tendency to pass from the sphere of language to that of eternal essences. The universal concepts which are a part of language and can be used legitimately as conceptual tools are themselves derived from the empirical sphere: because of the infinite diversity of the universe some general terms are needed to be able to communicate at all. The metaphysicians, however, are to be condemned for considering these general terms as corresponding to real essences, with an ontological status superior to that of single things. Nizolio proposes three definitions of knowledge: the ordinary knowledge of common things, the superior knowledge of difficult things which the vulgar are incapable of achieving, and knowledge of universals through demonstration. The first two Nizolio accepts, the third he forcefully rejects. Bruno's similar nominalistic tendency can thus be seen as part of a cultural development that existed throughout the sixteenth century.

Bruno's problem, however, was how to define a thing. When he claimed, in the *De umbris idearum*, a year after the publication of Sanches's text, that he was not repudiating Plato, even if he refused to accept his transcendental sphere of ideas, he surely meant that he was using a Platonic idealism to define the mode of being of single things. That is, all things in the universe are composed of two interrelated elements, the active idea (or form, or soul, or light), which acts from within like yeast forming the single objects of the phenomenological world, and the passive, material element composed of agglomerations of atoms. Because both of these aspects of the primal substance are fully infinite in spatial and temporal extension, there is no creation of Bruno's world and no transcendental God, although he does recognize an intelligent first cause which is logically prior to the infinite whole. The Platonic, and neo-Platonic, concept of a *descensus* from an intelligible sphere of being to the forms of sensible things, and a corresponding *ascensus* of the mind from the sensible to the intelligible plane of a full understanding, or true vision, remains intact in Bruno within the infinite universe of infinite vicissitudes. But now that there is no transcendental sphere of being to ascend to, understood in an ontological sense, the process becomes one of a purely epistemological ascent toward the full intelligibility of the single thing in its relations to other things. Like Nizolio, Bruno thus recognizes an ordinary, everyday kind of knowledge of things and also a higher and more difficult form of knowledge available only to the inquiring mind. The process can be presented (and at times Bruno uses this metaphor rather than that of ascent) as a penetration *into* the essential meaning of the infinite ocean of being. As in the other examples of this new sixteenth-century nominalism we have been discussing, knowledge for Bruno can never be a withdrawal or abstraction from the world of objects surrounding us.

We have so far ascertained, then, that Bruno's art of memory, although finding its sphere of inquiry in an infinite universe of being composed of an infinite aggregate of constantly moving atoms, is not concerned with knowing or defining either the infinity of the whole or the atomic minimum, both of which Bruno considers certain but unknowable. In section IV of Part II of the *De umbris idearum*, Bruno distinguishes between what he calls the "common ground" of the multiple subjects to be ordered by his art of memory and those multiple subjects themselves. The "common ground," he states, is excluded from his art of memory, which is concerned only with what that ground contains. The art of memory thus imposes order on finite things; although as the recent Italian translator of this text, Manuela Maddama, points out in her note to this passage:

> Once the infinite and homogeneous dimension is excluded, the remaining task will be to organize everything which, in its finiteness and disagregation, can be perceived by the limits of the senses, beginning with the known parts of the universe down to the primary mnemotechnical "atom." The expulsion of infinite space from the system generates a form of aggregation of the finites, and it is precisely by means of this autonomous aggregation of them at their various levels, followed by the starting up of a process of reciprocal combination, that it becomes almost possible to re-establish, through the finite and the multiple, the infinity and the homogeneity of the "common ground."[43]

Bruno points out in the passage in question that when he talks about the mnemotechnical "atom," he does not mean *atom* in the normal sense of the word but in the context of the art of memory. In attempting to explain this remark, it is important to note that Bruno is, once again, adopting some motives taken from Aristotle and some from Plato. From the *De anima* and the *Parva naturalia*, Bruno took the motive of the primacy of the imagination; Aristotle had insisted in both works that "without an image thinking is impossible." But Bruno would not have agreed with Aristotle that memory belongs only incidentally to the faculty of thought and essentially to the primary faculty of sense perception. The passage from the chaos of Anaxagoras to the ordered structure of the physical uni-

43. Bruno 1996:99; my translation.

verse is constituted for Bruno by the activity of the mind in the act of constructing ordered series of relationships with which to perceive and measure the order within the apparent chaos of the external world.

> This then is the method of discerning things, that the units (I might even say the many units, to concede something to the censors of words) which are being considered singularly become, by its efforts, disposed within an order. It is like the case in which, having marked a hundred sheep, each one with a different numerical symbol such as 1,2,3,4,5,6,7,8,9, and so on, those same sheep appear to us immediately afterwards as if confused in a disordered group, mixed up one with the other: so that then the shepherd with his stick separates the sheep, putting one next to the other: in the same way our cogitative faculty, with no hesitation, all other things being equal, chooses what has been determined in the proper order of succession according to its powers of discernment: . . . discernment then is a kind of number with which the cogitative faculty approaches in its way the images it has retained, dividing, dispersing, gathering, connecting, modifying, forming, ordering, relating them within the organization of the unitary form, all in proportion to its own capacity.

Luciana de Bernart has stressed the importance of Bruno's use of the traditional sheep image in the *De umbris idearum*, where it appears to define the theoretical basis of Bruno's art of memory.[44] The images the memory manipulates do not derive their original order from the experience of the senses, for an order—in this case the order of the natural numbers—has to be projected on to them by the mind to make the sensible world comprehensible. After working through this conceptual framework, which the mind has formulated as its instrument of research, it then approaches the apparent chaos of being, reelaborating, in time, patterns of order of an ever-increasing complexity and sophistication, in an ascending process of knowledge leading toward comprehension of the universal whole. This passage makes clear how for Bruno the process of true knowledge involves a use of mental parameters or ideas closely connected with the experience of the senses in order to achieve the increasing comprehension of things: mathematical or logical abstractions from the physical universe which is being subject to the process of knowledge described can only be a negative form of intellectual indulgence. This explains why Bruno was always so critical not only of Aristotelian logic, based on the concept of the universal abstract, but also of Ramus and his chains of reasoning which ignored the visual image, and, equally, of the

44. See the section *Organum*, pt. IV in Bruno 1991a and Bernart 1986; my translation.

modern mathematicians whom Bruno thought were engaged in mental juggling. On the other hand, if the mind is to be slipped like a glove over the ordered forms of being for knowledge to be valid at all, the dilemma of self-referentiality is already present in the way Bruno formulates its action. For the forms (numbers or any other ordered relationships) which the mind applies to comprehend the chaos of unformed being have already been supplied by the mind itself. What the mind "finds" in nature thus will be largely determined by the mental axioms and parameters it brings to its research. "And lo! the footprint is our own." Bruno's awareness of this problem is expressed eloquently in *The Ash Wednesday Supper*, where he complains that the painter of the universal picture (that is, the human mind) is unable to take a step back to contemplate the universe from a distance, as painters normally do to verify the perfections or imperfections of their composition, but has to contemplate it from within the picture itself: that is, within the coordinates of space and time.

I do not intend here to investigate how far the doctrine of memory elaborated by Bruno depends on the Art of Raymond Lull, mediated, as the critical tradition has abundantly established, through the "divine" Cusanus. Rather, I wish to emphasize the definition Bruno proposed of the mental parameters most appropriate for the search for natural truths as what he calls a "semimathematics." In his early memory and Lullian works Bruno uses mathematical figures, sometimes taken from Lull himself but sometimes invented, to define a mental geometry of association, rather than a geometry of pure forms in space. At this stage he is not so much interested in Euclidean geometry as in a geometry of the mind which defines the spaces and the movements of thought as it proceeds from subject to object, combining words, letters, numbers, or any other related images or signs into ever more subtle and refined patterns with which to trace the order dimly perceptible in the infinite chaos of the impressions derived from the senses. What the mind attempts to construct in time, using the images of an ordered process of memory, is a picture of the universal whole: a picture which, as Bruno repeatedly underlines, will be necessarily compressed, fragmentary, and incomplete. By constant attention to the behavior of the things dimly perceived, to the patterns of their life cycles, and to the necessities betrayed by their movements, the picture will be gradually refined and its accuracy improved. Thus the change in cosmology from the traditional Aristotelian-Ptolemaic system to the Copernican, and the extension of the Copernican system to Bruno's own spatial and temporal infinitism, represent improvements to the previous cosmological picture not because they comply to mathematical proofs but because the new picture takes into more complete and faithful account the necessity for all the worlds to revolve around central suns to obtain their heat and light, or the necessity of minds to proceed toward an intuition of the infinite.

In Bruno's early memory and Lullian works, patterns of numbers, patterns of words, patterns of various and often traditional images (the signs of the zodiac, the tarot cards) are all essential components of the pattern-forming activity of the mind. He can thus use to equal effect the apparently cold and formal symbols of the mathematicians and the apparently magically endowed images of traditional astrology or alchemy: the traditional uses of both are subsumed in their integration into the status of "linguistic" tools, or the formation of grids or ordered series of "events" through which to approach and trace the patterns within things themselves. Moreover, such languages are not incommensurable. As different examples of the pattern-forming activity of the mind, they can be translated one into the other: images can dissolve into words, words become numbers, numbers be replaced by letters or symbols. The isosceles triangle of Euclidean geometry can be described in terms of the doings of the Greek mythological gods: both are "alphabets" or "languages" which attempt an understanding of the vicissitudes of constant flux. It is their combinatory quality that interests Bruno and defines the efficacy of any one language or set: their capacity to set up a network of relationships stretching out through the ever-expanding sphere of being, or to contract into a primal sign or seal containing all being within it, like the Pythagorean tectratys or the temples of Apollo, Minerva, and Venus.

This search for images of total compression which contained all being within them had magical and theurgical origins and can best be seen in England in the work of John Dee, with his alchemical emblem, or *monas hieroglyphica*, and later in the cosmic images of Robert Fludd.[45] Bruno's formulation of the problem, however, seems to me essentially different from both Dee's and Fludd's because he sees the compressed images, such as the tectratys or the mathematical temples, as the dialectical opposites of images of an infinite universe, such as the infinite circle whose center is nowhere and whose circumference is everywhere. Both sets of images acquire their significance as pictorial "tools" with which to attempt to visualize the things in an infinite universe composed of atomic matter, rather than as magical talismans or icons containing within themselves some mysterious occult power.

It was, perhaps, just these traditional layers of powerful but ultimately

45. For the thought of John Dee, "Between Science and Religion," see Clulee 1988. For Fludd, see Goodwin in Hiley and Peat, eds., 1989.

indefinable meaning adhering to the imagery of the time which led Bruno to partly abandon the art of memory while he developed his atomic conception of matter. The nature of the minimum particle presents the fundamental problem behind his appeal to mathematics in *De triplici minimo*. Totally hidden underneath the species of the sensible universe, the minimum remains "unimaginable," or irreducible to an imagery that depends on sight and memory. In this case, and in this case alone, mathematics, with its conceptual abstractions, appeared to Bruno as possibly an appropriate guide. Once again, the mode of approaching the problem projects us forward to the dilemmas facing the physicists who, in the first decades of our own century, were attempting to come to terms with the minimum particles and their mysterious forms of behavior. Arthur Miller has discussed the revolutionary theories in physics of the first thirty years of the century as a problem of redefining scientific imagery. Miller underlines how the atomic theory of Niels Bohr, in the early 1920s, proposed a "picture" of the atom in which the movements of the particles making up the nucleus were seen in the light of planetary orbits, so that, as Max Born wrote, "the laws of the macrocosmos in the small reflect the terrestrial world." When the empirical data eroded the functional imagery of Bohr's solar-system atom, Werner Heisenberg was led to formulate the theory of quantum mechanics in which mathematics was to be the guide. In his historic paper of 1925, Miller narrates, Heisenberg emphasized the failure of our "customary intuition" when faced by quantities of atomic dimensions and underlined the necessity for particle physicists to liberate themselves from "intuitive pictures." This, however, proved far more difficult than Heisenberg anticipated; and the early development of quantum mechanics was closely tied up with a discussion between Bohr and Heisenberg on the ways in which particle physics necessitated a departure from visualization in the usual sense. In a letter to Wolfgang Pauli of May 16, 1927, Heisenberg wrote: "There are presently between Bohr and myself differences of opinion on the word *anschaulich*."[46]

It is not, I believe, unhistorical to link Bruno's mathematical dilemma and his intense preoccupation with problems of mental imagery to such modern developments in physics. The problem has haunted the philosophical and scientific endeavor from the beginning. Its roots lie in Platonic and Aristotelian concepts of the mental image. If the emphasis on sense impressions on the part of the seventeenth- and eighteenth-century empiricists posed the problem of mental imagery in relatively simple terms, George Berkeley's new idealism proposed the more problematical

46. See Miller 1984, esp. chap. 6.

possibility of imagining something that cannot be perceived, for example, abstract images of things. Kant placed the problem on new ground by giving a central place to the imagination with its mysterious elaboration of mental pictures within the mental categories of space and time. Although Kant apparently did not read Bruno, there is now abundant evidence that others did read Bruno's works on memory and mental imagery in the directly post-Kantian years. Rita Sturlese, in her introduction to the *De umbris idearum*, provides this historical perspective:

> The hour of re-discovery of the *De umbris* arrived only in 1800, in Germany, clearly in the wake of the renewed historical perspectives opened up by the new critical idealism: in the *Geschichte der neuern Philosophie* the Kantian Buhle decided to break with a centuries-long tradition and to offer nothing less than a subtle, detailed and systematic exposition, underlining how Bruno's text not only contains a complex set of mnemonic rules, but also an analysis of the entire spontaneous activity of the mind in the act of thought. It is on the pages of Buhle that Hegel would found his interpretation of Bruno's art of memory in his *Lectures on the History of Philosophy*, presenting it as "an attempt to comprehend the universe in its development, in the system of its determinations, and to demonstrate how the outside world is a sign of the ideas."[47]

If Bruno's mathematical doctrine was above all developed as a function of his atomic theory of matter, it was destined rapidly to reach an impasse. Looking ahead, as he was, to the moment of discovery of minimum particles and to the complex problems involved in measuring and defining them, he could only abandon the subject to a later era and return, as he did in the *De imaginum, signorum et idearum compositione*, to the general art of images and memory. Hegel's interpretation of the value of his works on this subject offers the lead with which to consider them in the context of Bruno's science. These works, which culminate with the *De imaginum*, may be seen as an attempt to offer a scientific treatment of the mental image in terms of the complex functioning of the mind in time and space. In this they are quite different from the renaissance emblem books, which supply dictionary definitions of the various images and their traditional meanings. Bruno used these definitions, but they were not the principal substance of his own texts. Nor were they the kind of memory works envisaged by Camillo and satirized by Citolini, which attempted a macroscopic, encyclopedic systematization of memory images on a cosmic scale. They were, as Hegel realized, properly philosoph-

47. See the introduction to Bruno 1991a; my translation.

ical investigations of the image-making properties of the mind, tending toward a scientific form of systematization of the various moments of imaginative thought.

The *De imaginum* starts with a theoretical investigation of images in the mind which, as Bruno himself admits, treats the problem as already elaborated in the early *De umbris idearum*. He even hints that his new text had actually been written some years previously. The three "moments" of being which define an object remain the same: there are the ideas of single things which are prior to the things themselves; these ideas produce the things which are always composed of some combination of matter with form; shadows of the ideas, which are the mental species, appear in the mind, which can remember, combine, and elaborate them into new and vital forms.[48] It soon becomes clear, however, that these three moments are not really separated or separable but tend to coalesce in the image-forming activity of the mind which is subject to time and space and memory. Images and signs of things derived from the senses exist in the mind together with those of intelligible ones. Some images and signs are in things, but others are in intentions, in words, in graphic delineation, in the sounds of voices. In a chapter concerning the images of things, Bruno calls these primary images derived from objects "certain fixed and tiny elements," rather like mental atoms. These sound a bit like the primary sense perceptions of Locke, or perhaps even more like the *petites perceptions* of Leibniz.[49] Then Bruno introduces the Platonic idea of the mind as a tablet of wax, which immediately begins a process of subtle metamorphosis, introducing conceptual elements that form order out of the primary perceptions. Bruno finishes up by conceding preeminence to these principles of the mind. For although we receive a forest of chaotic impressions from objects at one glance, the mind allows us to contract and judge them in an instant.

This mental ordering power is considered by Bruno to be a form of light: not, however, ordinary material light, like that created on the fourth day of Genesis, but the divine light of the first day. Again, Bruno associates this mental light not only with the Old Testament account of creation but with the account of the creation narrated by Hermes Trismegistus in the

48. See in particular Spruit 1988 and 1995:203–13.

49. For Bruno's text, see *De imaginum, signorum et idearum compositione*, bk. I, pt. 1, chap. 7: "Concerning the images of things" in Bruno 1879–91, II, iii:105–6, and Bruno 1991b:21. For the importance in Leibniz of the *petites perceptions*, see the Preface to *New Essays*: "insensible perceptions are as important in the philosophy of mind as insensible corpuscles are in natural sciences," quoted and commented on in Jolley 1990:162. See also Mates 1989.

Pimander. Both these stories are clearly to be considered as metaphorical, for Bruno makes it abundantly clear that his infinite universe is eternal and not subject to a moment of creation in time. What interests him both about the biblical account of creation and the Hermetic one is the emphasis on this metaphysical light. It is the element present throughout the whole infinite universe which combines ideas, objects, and mental images into one unified "moment" of illumination. Without it, ordered understanding would not be possible.[50] Recalling the doctrine of the light of the soul propounded by Saint Augustine, Bruno here appears to anticipate a central theme of the later Cartesian doctrine of the mind. Nicholas Jolley has pointed out that "for Augustine divine illumination can be no more than an account of our knowledge of necessary truths. The Cartesian revolution in philosophy provides an opportunity to transform the theory of divine illumination into something more radical; with the aid of Descartes's discoveries, it can be incorporated in a general theory of perception."[51]

Having thus established his theoretical bases for mental imagery, Bruno goes on to attempt to give an account of how various types and classes of images are formulated in the mind, to discover the unifying criteria that underlie them, and to demonstrate the ways in which an art of memory can be established to remember and manipulate letters and words. A notable aspect of his treatment of the theme of images is the idea that language is fundamentally coexistent with the presence of images in the mind. I am unable to find in Bruno the Platonic conception, expressed in a famous page of the *Phaedrus*, of the written language as a negative form of distortion or deceitful sophistication of primitive understanding. Bruno seems rather to consider words, translated into signs, as coeternal with images. He is interested in early forms of sign making, such as the hieroglyphs that so fascinated the renaissance; but he does not seem to me to think them superior to modern forms of linguistic communication. On the contrary, the chariot depicted as a chariot seems to him too crude a way to make a sign, and he approves of the Egyptians when they discovered more complex forms of sign making. Nor, I think, is he disapproving of the Greeks when he calls them "subtle" with words, imputing to them the invention of the metaphor, an important linguistic instrument for Bruno because of its powers of relativization of concepts.[52] What

50. For Bruno, see *De imaginum*, bk. I, pt. 1, chap. 12, in Bruno 1879–91, II, iii:117–18, and "Of light, the vehicle of images" in Bruno 1991b:36–37.

51. Jolley 1990:92–98.

52. See Bruno's *De imaginum*, bk. I, pt. 1, chap. 9: "Some ways of fashioning and deriving images that Chaldean writing holds in Mnemosyne's temple" in Bruno 1991b:28–

he disapproves of are the Latin-speaking pedants of his own day who insist on repeating the traditional metaphors found in the texts of antiquity. All periods must find their own metaphors, in Bruno's opinion, for the metaphors of each age must correspond to their own changed reality. "Just standing still changes the names around me."

Bruno's treatment of language, or what he calls "the images of words," is both analytic and sophisticated. In the lively renaissance debate, centered on Plato's *Cratylus*, about whether language was composed of conventional signs or whether words were intrinsically linked to the essence of the things they depicted, Bruno appears to me to assume the conventionalist position.[53] A system of signs may begin unconsciously but in a cultivated society should be conceived logically as "in advance," or predetermined, so that it can function adequately as a system of notation. If single letters are conceived of as being "like" certain things, for example the O like a sphere, or the A like a ladder or compass, this should be recognized as simply an enhancement of a text. Words can nevertheless grow from these "likenesses," acquiring numerous syllables in organic fashion, finding their own ramifications and grammatical structures according to combinations of the original elements, or letters, of whatever "alphabet" is concerned. Bruno believes that languages derive from their primal "seeds" rather as human bodies do. He even maintains that a grammar can be understood in terms of a human body: the nominative case being the head, the dative the right hand, the genitive the genitals, the vocative the mouth, and so on. He knows the categories of conventional linguistic analysis, such as metalepsis, division, contraction, etymology, and interpretation. But he has added to these, taking into particular consideration the problem of heads and tails of words: that is, their first and last syllables and letters. He maintains that words can form meaningful strings by being treated in these terms: for example, one can go straight from "ass" (*asinus*) to "asylum" (*asylum*), or from "temple" (*templum*) to "contemplation" (*contemplatio*). Echoes and resonances, Bruno seems to be saying, are set up between words which go beyond grammatical rules; and one remembers that he was also a poet.[54] Indeed, the extent to which

---

30. Plato in the *Phaedo*, 74a, considered the metaphorical capacities of the mind as dependent on an innate idea of our imperfection with respect to God, a concept which, through the *Itinerarium mentis in Deum* of St. Bonaventura, will be taken up by Descartes: see Cottingham 1993:145–66. Bruno's opinion on this subject is closer to that of Aristotle, who, in the *Rhetoric*, bk. III, 1407a 10–17, writes of "proportional metaphor" which must always apply reciprocally to either of its coordinate terms.

53. The importance of this debate for the birth of a new science has been underlined in Vickers ed., 1984:95–163.

54. See *De imaginum*, bk. I, pt. 1, chap. 10, in Bruno 1991b:31.

Bruno's own poetry, and especially his use of the sonnet in *De gli eroici furori*, should be connected to the same passion for the making of order which underlies his art of memory is a question still to be defined by the critical tradition.

The largest part of *De imaginum*, as the title indicates, deals with imagery. Bruno clearly respected traditional doctrines and cultural assumptions which he realized acted as cohesive factors in a society profoundly lacerated by the wars of religion he hated so intensely. In the *De monade*, the second work of his Frankfurt trilogy, he returned to Pythagorean number symbolism, investigating, as he put it in the dedicatory letter to the duke of Brunswick, "the way in which revelation, faith, and divination constitute the foundations or shapes of imaginings, opinions, and experience." His investigation of these commonly held doctrines associated with numbers had not, he specifies again in the dedicatory letter, led him to completely satisfactory results: the research had been carried out "not without uncertainties." Nevertheless, Bruno appears to have thought that imagery, even more than numbers or words, arises out of a collective unconscious. He turns his attention once again in the *De imaginum* to the ancient and traditional sources of imagery and finds himself deeply involved with the imagery of astrology. The English translators of the *De imaginum*, the first of Bruno's Latin works to be translated into English, have found the source of these images in the *Flores Astrologici* of Albumasar (Abu Ma'shar), the ninth-century Moslem astronomer. A printed copy of Albumasar's work had appeared in Venice in 1500; but the editors of the *De imaginum* think that Bruno's direct source was the 1563 edition of the works of the Venerable Bede, *Opera Bedae Venerabilis Presbyteri*, published in Basel in 1563, where they appeared as illustrations of Bede's "Mundi Sphaera."[55] Bruno knows all about the conceptual foundations of astrology; as cultured people did in his day. He illustrates this knowledge in one of his most interesting chapters by recognizing that the forms, simulacra, and signs are vehicles, or chains, by which the favors of lofty things emanate from above and are drawn down into the lower order. Potent effects, Bruno writes, are thought to originate in celestial bodies, which are considered to possess virtues and to irradiate them at particular moments of the day or the year. This is the reason for certain magical incantations, or for the crosses and charms that can be seen at crossroads: attempts by people to protect themselves by figures, to defend themselves from their enemies, and to gather down for their own benefit the powers of the stars. It would, however, be a mistake to identify Bruno himself

55. Ibid., xxxix–xiv.

with such practices. He recognizes their existence and their power in the society of the time, but he refers to them as "modes of general reception" rather than as representing his own convictions. At the end of this same chapter he appeals to what he calls the most "adept minds," defying the censure of the supercilious and the ignorant, to claim that the light of preordained principles in the mind, not external astral powers, ensures the efficacy of the truths derived from physical, mathematical, and logical images. He then backs up this observation by deploring those modes of thought which descend from universal truths too broad for direct comprehension. The strong power of the image derives from its status as a particular definition linked to the object represented, starting from which it is permissible to proceed to the next genus and go on from there. What is not permitted is to start from the remote and the primal. Here, surely, is a page Bacon may have read.[56]

Images, then, like languages in words, start from simple forms like fixed, atomic elements and then coalesce into more and more complex aggregations. Painters tend to use their imagination to manipulate extraordinarily complex images that radiate out from the center to the fullness of eternity "where we mostly live." Bruno himself, however, stresses the starting point of this process, which has to be clearly recognized and investigated: the "places" of memory in each single mind where the primary elements are first reduced to order. Without the places or "halls" of memory, consecutive thought could not exist. Bruno illustrates these halls as part of the classical art of memory, with pages of complicated tables and groupings of words and images that generate new words with their accompanying imagery. What Bruno is illustrating is not astral magic but mental magic: the formal process that generates words and images in time, one from another, at the origins of that remarkable process we call thought.

Bruno's purpose, in my opinion, in his works on memory is to formulate an account of the processes of thought which is different from an abstract logic. He attempts to illustrate the ways in which the primal chaos of impressions is reduced to order by principles innate to the mind; at the same time he takes into account the historical and social processes through which languages, both of words and images, have developed organically during the course of civilization. The image of the tree to signify the mode of growth of languages, derived from Raymond Lull, acquires in Bruno a historical dimension. The social consensus is seen to be

56. *De imaginum*, bk. I, pt. 1, chap. 5. See Bruno 1879–91, II, iii:101–4, and Bruno 1991b:16–19.

an important part of what is considered truth; for the ways in which, at any time, words and images are used depend not only on the power of imagination of the individual but also on the shared conventions of the society in which he lives. This awareness of the historical growth of languages and imagery tends to limit the possibility of applying his works on memory to the dramatic needs of the new science to develop a new logic of inquiry. Bruno seems to have seen the development of a new mathematics as negative, separating the new investigations of the universe from traditional conceptions, such as number symbolism, which were still in Bruno's time part of the social and communal modes of apprehension of the real. Yet Bruno was aware of the necessity to create new languages and images to describe the changing universe being discovered by the new science. In the first chapter of the first book of *De triplici minimo*, he insists that a new beginning will be made only when the old words are torn up from the deep shadows of antiquity and new words are invented in accordance with the new scientific doctrines. Here he comes close to Bacon's famous later repudiation of the traditional mental "idols" that uselessly cluttered up the scientific mind. In the *De imaginum*, however, Bruno appears unwilling to abandon traditional forms of thought and language quite as ruthlessly as he himself in *De triplici minimo*, and Bacon after him, prescribed.[57]

It would be only too easy, in a book dedicated to Bruno and the scientific revolution of his time, to dismiss his study of imagery as conservative and backward-looking, one of the factors that limited him from contributing effectively to the growing scientific activity of the end of the sixteenth century. I believe that would be too simple a conclusion. Bruno's difficulties with the new mathematics were certainly based on a conviction that it offered too abstract and schematic a picture of reality. The remaining epicycles and eccentrics of Copernican astronomy, and the approximations of the trigonometers, were, in Bruno's opinion, untruths. Greek mythology might have given a truer picture; at least, it would be wise not to throw out mythology while hanging on to Euclid. Bruno thus can be seen as insisting on the primacy of a natural universe, of its mysterious affections and vicissitudes, which he understood that the new science, with its abstractions and specialized languages, would manipulate and often disregard. There is more to Bruno's historical dilemma than

57. For Bacon's treatment of the four idols see the *Novum Organon*, aphorisms xxxviii–xliv, in Bacon 1858–68:250–53. Brian Vickers 1968 has claimed, however, that Bacon himself was much more closely attached to traditional renaissance rhetoric than his page on the "idols" would lead one to suppose.

this, however. He constantly returns to the problem of how to understand the infinite complexities of those vicissitudes with the powers of a finite mind, limited to a fixed position in space and time. To Bruno it clearly seemed a desperate enterprise when the primal element of being, the atom, was impossible to see, let alone to measure and understand. There were also dangers on all sides, from the oversimplifications of the new scientists, with their exaggerated abstractions, to the "living with eternity" of the theologians, which Bruno considered an illusion given the finite dimensions of the human mind.

For Bruno the greatest problem of all was to be found in the mind itself, constrained to operate in categories of space and time which might not prove adequate for a comprehension of the infinite whole. One way Bruno expresses this dilemma is to say that the mind is like a painter creating a picture without being able to stand back from the canvas to judge how faithful the picture really is. Another way is to say that the eye sees an outside world but is unable to see itself.[58] In either case, the operation of the mind in the process of scientific discovery is the problem which Bruno is already posing with anxiety and force. Realizing that the mind projects its own principles on to the phenomenological world with which it finds itself in such a close relationship, he was aware of the danger of the "loop" of self-referentiality. Like Roger Penrose and many others writing about the scientific enterprise today, Bruno concludes that if our concepts of the natural universe are ever to be properly understood, we must pay particular attention to the nature of the minds investigating it.[59] It seems to have been in the light of such concerns that Bruno turned back, in the last months of freedom before imprisonment, to the renaissance doctrines of magic, concerning himself with the traditional magical concept of "links" between differing grades of being but transposing that concept from a metaphysical sphere to a study of the links between mind and mind.[60]

58. These remarks on the art of the painter, so essential for a correct understanding of the cosmological dialogue which follows, are to be found in the *Proemiale epistola* of *The Ash Wednesday Supper*.

59. See the final section of Penrose 1989. Also Morowitz 1981:34–49. Writing of the crisis period of the 1920s, which saw the invention of quantum mechanics, Morowitz writes of the German physicist Werner Heisenberg that he "stressed that the laws of nature no longer dealt with elementary particles, but with our knowledge of these particles—*that is, with the contents of our minds.*"

60. For a study of Bruno's transposition of the traditional renaissance magical discourse from a metaphysical to a psychological plane, see Cambi 1993:9–33.

What Bruno was attempting, then, in his art of memory, was a philosophical investigation into the image-making properties of the mind added to an attempt to propose a picture logic sufficiently flexible in its powers of association to act as a guide, in time and space, through the intricate finite vicissitudes of a newly atomic and infinite universe. As Stephen Clucas writes: "The artificial memory which he [Bruno] outlines in his mnemonic works is a twofold art (*artem . . . sub duplici forma*). First it teaches a theory of the soul, outlining the 'ordination of all the soul's operations' (*omnes animi operationes ordinandos*), secondly it provides a practice: that of artificially cultivating memory and using it for the purpose of invention."[61] Behind this attempt lies a question Bruno never explicitly poses: whether a picture logic could supply the indisputable logical proofs the new science needed so urgently to convince its many detractors. Alexander Dicson's Ramist antagonist from Cambridge, William Perkins, evidently thought not. So did Galileo, who complained in his *Dialogue* of the traditional number system, claiming that even William Gilbert, the magnetic philosopher he so admired, had failed to understand the primary importance of a truly mathematical physics. It was in these mathematical terms that the new science would solve its epistemological problem until the classical mathematics began to lose its dominating influence at the end of the nineteenth century. In the context of this more modern crisis, it is of interest to notice that one of the problems posed by philosophers today, in a society once again dominated by the visual image, is whether, and to what extent, images can be considered to mark developing processes of thought that culminate in logical forms of proof.[62]

Bruno never succeeded in creating a system of mnemonic images or signs capable of providing new answers about the infinite, atomically constructed universe he envisages, thus obliging him, in the more scientific parts of his discourse, to fall back on a mythologized version of Euclidean geometry. Nevertheless, his attempt to approach the universe of infinite things through ever more complex systems or "programs" of connected images or signs foresaw the inevitable crisis of the classical mathematical physics which was developing in his time and has implications for the computerized science of our own day.

61. See Clucas 1997.
62. See Giaquinto 1992, 1993, and 1994.

11 Alienation and Reconciliation

My discussion of Bruno's epistemology, as it moved questioningly and uncertainly between mathematics and a picture logic of memory, has attempted above all to relate it to his crisis ontology defined in terms of an infinite universe and an atomic conception of matter. I found myself referring with increasing frequency, however, to the lively, ongoing epistemological discussion of our own times, which is involving ever more intensely not only philosophers and philosophers of science but also practical scientists from an increasing number of disciplines. Wolfgang Pauli's essay of 1952, "The Influence of Archetypal Ideas on the Scientific Theories of Kepler," which compared Kepler's geometrical imagery and the magical-occult cosmological imagery of Robert Fludd, was the first to relate the crisis in visualization caused by modern quantum theory to discussions about scientific imagery conducted at the origins of the scientific revolution.[1] Since then, an increasing number of contributions to the discussion of the epistemological crisis caused by relativity theory and quantum mechanics have referred to late renaissance thinkers who seem to suggest modes of resolution of what many years ago, in a literary context, T. S. Eliot called the modern "dissociation of sensibility."[2]

Eliot was concerned with the same phenomenon that interested Pauli:

1. Garin 1961:143 called attention to the paper by Pauli; it has also been the subject of a major contribution by Robert Westman (1984:177–229).

2. Eliot's phrase was first introduced in *The Metaphysical Poets* published in 1921. It became a favorite tool that he used in later critical essays, particularly in the much-discussed critical treatment of John Milton in an essay of 1936.

the triumph of rational, linguistic discourse at the end of the sixteenth century, on the Baconian model, with the consequent loss of visual imagery. Eliot, obviously, approached the problem from the point of view of a poet, and he thought the major cause of the crisis in sensibility was not so much the new science as the Protestant Reformation with its hatred of images and icons. Frances Yates has stressed the importance of the Reformation in the loss of visual imagery in this period in her discussion of the Dicson-Perkins controversy and the emergence in England of Ramian logic with its consecutive, nonpictorial chains of reasoning.[3] Behind both Eliot and Yates can be sensed the presence of the work of Max Weber and his much-discussed essential link between the Protestant Reformation and the industrial society and economy.[4] Robert Merton in 1938 proposed a thesis that claimed a further link between Protestantism and the institutionalization of the new science on which the industrial revolution was based, added to an increasingly complex historical picture in which a combination of factors created the dissociation between body and mind, image and thought, which lay behind the development of a rational, empirical science.[5] More recently philosophers have been pointing at Descartes, rather than at Protestantism or Ramus, as the originator of the mind-body dissociation which is at the center of so much current discussion.[6] In spite of this variety in the deciphering of causes, however, there appears to be increasing agreement on the necessity for science, and indeed for culture generally, to effect a recomposition of what is increasingly seen as a split psyche. From Richard Rorty's repudiation of the mind as a mere "mirror of nature," to the conviction of Roger Penrose that deeper understanding of our modes of consciousness must complement a

3. See Yates 1966:260–78.

4. First published in German in 1904–5, Weber's *Protestant Ethic and the Spirit of Capitalism* was translated into English by Talcott Parsons (New York, 1930).

5. Merton's thesis was proposed in chaps. 4–7 of *Science, Technology, and Society in Seventeenth-Century England* (New York, 1938) and discussed again in a new preface to the 1970 edition of his book. For a reply to Merton's many critics claiming that his thesis must be understood as primarily sociological rather than historical or epistemological, see Abraham 1983.

6. The anti-Cartesian attack was already explicit in the final section of Yates 1964: 453–54. In the form of a repudiation of epistemology as such, or a foundational "theory of knowledge" as the basis of philosophical discourse, it has come primarily from America in the work of Sellars, Quine, and Richard Rorty. For a summary of their view, see the introduction to Rorty 1980. This tendency involves the claim made by David Davidson that the theory of meaning is going to be empirical rather than epistemological, as well as the claim by Hilary Putnam (1975) that the scientist's story "leads to no reason to think that scientific theory is true or that scientific practice tends to discover truth."

science of purely computerized predictions, to the "implicate order" of David Bohm in which being tends again to coincide with meaning and not simply with a mathematical succession of probabilities, voices at the center of discussion of modern science appeal to an ideal of objective cognition more closely related to the subject, more humane and complete.[7] The fragmentations and interruptions in the images of the real offered by quantum mechanics; the obsession with single, clearly defined events at the expense of a larger picture of the whole; the increasing uncertainty of the precise ontological status of what is being observed: all these themes again and again signal that science itself is becoming increasingly aware of its epistemological shortcomings. In a contribution to a volume in honor of David Bohm, B. C. Goodwin has suggested that modern biology rather than quantum mechanics or computer technology should be taken as the model of scientific knowledge, supplying as it does a "science of union" rather than of fragmentation.[8] Even in that field, however, after the discovery of DNA, RNA, and their interactions, ingenious techniques of molecular analysis have yielded, in Goodwin's opinion, remarkable descriptions of the processes involved in reproduction and evolution while remaining elusive at a higher level: "namely, the nature of the integrated spatial and temporal order that gives organisms their distinctive attributes, particularly their morphology and their behaviour."[9]

Goodwin's essay discusses the present-day problem of complementing a purely quantitative science with a science of qualities in the light of the Yates thesis, developed around the "unifying themes" of the renaissance magi. Goodwin (explicitly following Yates) sees the "mystic natural philosophy" of Ramon Lull as inspiring a long line of renaissance thinkers devoted, perhaps a little vaguely, to "unified and unifying change." Francesco Giorgi, Johannes Reuchlin, Giordano Bruno, and John Dee are bunched together under the heading of "magi" devoted to this "science of

7. See Rorty 1980, the final section of Penrose 1989, and Bohm 1980.

8. See the relevant essay in Hiley and Peat, eds., 1989. This high estimation of modern biology and botany, and a concomitant downgrading of modern physics, is apparently shared by contemporary physicists. H. R. Post writes that "electromagnetism, for instance, is unable to answer the simplest empirical question meaningful in its field of application, namely the question as to the motion of an electron in a uniform electrostatic field. On the other hand, botany has fair predictive power in its own field of application" (1993:7, n. 14). This statement is quoted approvingly by Feyerabend 1975:61, n. 17, in support of his own thesis that Aristotelian science might have been more adequate to its task than the more abstract sciences that replaced it. Feyerabend, however, disagrees with Post's rationalistic and pro-Popperian stand on scientific theory.

9. See Hiley and Peat, eds., 1989:328–37.

union," as if there was little to distinguish the various members of the group. In this Goodwin is, of course, only following Yates; and I wish to take the opportunity of commenting on his essay to make clear my own position with respect to the various aspects of what has become known as the "Yates thesis."

Those who use the convenient catchphrase, the so-called Yates thesis, do not always sufficiently emphasize that it actually has two distinct elements. The first of these, chronologically, was the interpretation of Bruno himself as a particularly radical and extreme example of what she called a "Renaissance Magus": that is, a philosopher imbued with renaissance magical doctrine based on Ficinian neo-Platonism and Hermeticism, complemented by Pico della Mirandola's Christian cabalism. Bruno, in Yates's opinion, had long been mistaken as "an advanced thinker, heralding the new cosmology which was to be the outcome of the scientific revolution. . . . The philosophy of Giordano Bruno, instead of being studied as has been done in the past in isolation from its true historical context, can now be examined by historians of thought as a remarkably complete example of a Hermetic world view in the immediately pre-scientific age."[10] From this interpretation of Bruno there emerged gradually a larger thesis which claimed a complex and at times rather uncertainly defined relationship between the newly emerging science and the long tradition of renaissance magi to which Bruno, in Yates's work, is assimilated.

A detailed discussion of the further development of this larger thesis must, in my opinion, take into account that Yates herself was not always completely consistent in her comments on the fertile and important question which she had so persuasively raised concerning the relationship between the occult and the scientific traditions at the beginning of the so-called scientific revolution. At times she maintains her idea of the fundamental incommensurability between the two traditions and their diverging mentalities, claiming that the only real contribution of the magi to the birth of a new science was their desire to dominate and manipulate nature. At other times she underlines how all the new scientists, even those usually considered pure rationalists, such as Bacon or even Newton, were still deeply involved with elements of alchemical, Hermetic, and occult culture. In this case, she posed the rather different question of how the one strand eventually freed itself from the other. The research into renaissance scientists and science of the last thirty years has proved this wider aspect of her thesis to have some foundation; even if

10. Yates 1964:451.

at times it has been overstated both by Yates herself and her followers. Nevertheless, the question she posed is one around which much discussion of the scientific revolution has been moving.[11] But if this larger dimension of the Yates thesis was more complicated and at times more contradictory than is always appreciated, her interpretation of Bruno never wavered. He remained for her until the end the supreme example of the renaissance Hermetic Magus. The English-speaking world has almost unquestioningly accepted him as such for the past thirty years, endorsing, in my opinion, a reductive reading both of his life story and of his works.

I have tried to show in this book how Bruno's infinite cosmology led inevitably to an ontology that precluded identifying his epistemology with that of the renaissance magi. From Marsilio Ficino to Robert Fludd, the renaissance neo-Platonists accepted a hierarchy of being, and generally speaking also a geocentric or at least a one-world cosmology, which differentiated their discourse radically from the philosophy developed by Bruno. There were certainly times, as I have repeatedly underlined, when Bruno addressed their doctrine of being, their prevailing imagery, and their analogical mode of reasoning, comparing them, at times favorably and at times unfavorably, with the new, more rationalistic modes of discourse which were emerging in his time. But Yates constantly fails to distinguish those pages in which Bruno is summarizing their doctrine from those in which he is criticizing or endorsing it. In her reading of him, any reference on his part to the renaissance magi amounts to simple identification with their doctrine. The result has been to deny him any contribution to the emergence of the new scientific discourse that was at the center of his attention throughout his philosophical life.

I will conclude my treatment of Bruno's epistemology by underlining my suggestion that his thought can usefully be considered in the light of the present-day discussion of the crisis in modern science, with its desire for the formulation of doctrines capable of complementing the remarkable techniques of calculatory analysis which have recently been achieved. It would, however, according to my reading, be erroneous to see him as one of a line of renaissance magi whose interest (like that of Fludd for Pauli) lies in their quite different and still essentially nonrationalistic approach to an investigation of the natural world. The interest and originality of Bruno's philosophy (and the argument could be extended to include other renaissance natural philosophers such as Cardano, Telesius, Francesco Patrizi, and Tommaso Campanella) lies rather in his awareness

11. The start of this discussion came with the much-read article by McGuire and Rattansi 1966. For its further development, see Copenhaver 1990:261–301.

of a crisis that was emerging with the advent of the new rationalism: a crisis in that antique unity of knowing and being, mind and body, subject and object, which found its linguistic dimension in a sense of identity between names, or images, and things themselves. The realization of a new sense of alienation between the human subject or mind and the universe seen as its natural environment endowed with meaning by God haunts much of the most sophisticated natural philosophy of this period, complementing the enthusiasm and optimism of the new rationalistic and empirical scientists. This properly philosophical and partly skeptical attitude is nevertheless to be distinguished from the attitude proper to the Magus, who was also, in his very different way, optimistic about his capacity to know and act within a reality dominated by spiritual essences and correspondences whose secret relationships could be deciphered by magical techniques. Whether the secret of the universe lay in a Ficinian magical chant, or, as Dee thought, in a magical symbol like his Monas Hieroglyphica, or as Fludd was convinced, in the mysterious behavior of a grain of wheat, the magician was confident, provided the evil forces of deceit and hellish manipulation could be avoided, of his powers to capture these resonances in his mind. The Magus held a "philosophical key" in his charmed hand, which he believed allowed him to enter the multiple doors of being.

Bruno understood the attraction of magic, which, in *The Expulsion of the Triumphant Beast*, he defines as a precise mental "attitude" and, following a classification already rendered canonical by Cornelius Agrippa, divides into three kinds: divine, natural, and mathematical. But his treatment of the problem is historical, associated with a consideration of ancient Egypt and the figure of Isis. Even the famous celebration of the ancient Egyptian religion at the end of the same dialogue is historical in tone, although at times assuming a celebratory rhythm that betrays an undoubtedly sympathetic attitude. The passage is largely elegiacal; however, for, as the words of Smitho at the end of Dialogue 1 of *The Ash Wednesday Supper*—ultimately, if perhaps reluctantly, endorsed by Theophilus—had made clear, the serpentine folds of history had made the times of the "old philosophers" stories of long ago. There is never any illusion in Bruno's philosophical works that it was possible, at the end of the sixteenth century, to go back to ancient pictures of the universe, even when they interested him for the closeness they portrayed between the mind and the surrounding universe. Bruno knew that his culture was going forward very rapidly. His concern was with the direction it was going in, which seemed to him to open up new and undreamed of vistas of knowledge but at the same time to spell out new forms of alienation between the mind and the world of objects of which it was becoming increasingly aware. His crisis was the same as the one that worries modern philosophers and scientists, at the beginning rather than the end of an extraordinary period of scientific investigation. Rather than read Bruno as some opposite spirit, interesting perhaps precisely because not yet aware of or involved in the modern dissociation of mind and body, subject and object, it is more useful to read him as a philosopher already addressing the same concerns, although still in an embryonic form, as those worrying postrelativity and postquantum scientists and philosophers. I will illustrate this claim, and conclude my discussion of Bruno's epistemology, by a reading of two passages from the final Frankfurt trilogy: Bruno's treatment of the Pythagorean tectratys in the *De monade*, and the final three chapters, 8 to 10, of book VIII of his final cosmological treatise, the *De immenso et innumerabilibus*.

In the *De monade*, Bruno approaches his treatment of the Pythagorean celebration of the number 4 through meditation on the number 3.[12] After recalling the various traditional meanings attached to the number 3, in his most interesting section of this chapter he treats 3 as the number representing the logical movement of the mind: the number sacred to intellectual being. Mind thus has the form of a triangle, which Bruno places in a circle called the ring of Apollo. The circle symbolizes the eternal perfection of the divine intelligence, source and fountain of the infinite whole. The equilateral triangle situated within this circle represents the triadic movement of logical thought. Although Bruno was well versed in the Aristotelian logic of the syllogism, his triangle is not designed to illustrate any form of specific syllogistic reasoning. Rather, the various points of his mental triangle represent three entirely different stages of consciousness and as such are closer to Hegelian dialectic than to Aristotelian logic. The first point of the triangle marks the intuitive apprehension of the whole; the second, the rational comprehension of the simple elements in their dualistic tension; the third, the moment of synthesis which Bruno thinks of as a moment of action or speech. Each of these three moments of consciousness can be represented as a circle within a larger triangle of the mind. Bruno thus draws an illustration of a triangle containing within it three equal circles which touch each other, while the sides of the triangle act as tangents to the three circles. He calls this the image of the Three Graces. It is this image which leads to the seal of the *tectratys*: the Pythagorean symbol of the number 4, which contains within it all form and number; for adding together the natural numbers

12. See Bruno 1879–91, I, ii:358–75.

up to and including 4 gives the number 10, which represents the entirety of the decade. Bruno uses this symbol to indicate what he calls, with echoes that are both Epicurean and Pythagorean, "the ocean of being." For Bruno the *tectratys* defined in these terms is closely connected with the triangle; for from all triangles it is possible to derive all forms of parallelogram. In Bruno's opinion, this concept shows how many things happen by consequence. The triangle of the Three Graces thus illustrates the mental category of cause and effect followed by action and teaches us how to cross the ocean of being.

His new inquiries, then, will be concerned with a search for causes and effects which will finally free the mind from its long centuries of contemplation on the shores of being and allow it to strike out across the ocean itself. In the final chapter of the *De immenso*, Bruno restates this theme in terms of the metaphysics of light, which would be an important metaphor for the new ages of rationalistic inquiry.[13] Bruno's metaphor of illumination, however, has characteristics that differentiate it radically from many of the light metaphors being developed in his time. In particular, as we have seen, Bruno criticizes Palingenius for dividing his double concept of light into a divine light of superior being, lying outside the sphere of the fixed stars, and a lower form of light proper to the elemental world. Bruno's infinite substance is throughout imbued with light but partakes equally of its opposite, which is shade. These two factors represent for Bruno the two aspects of being: its material element and its inseparable formal or life-giving element, which is its light. There is no light "flowing from above," for there is really no "above" in Bruno's infinite universe for it to emanate from. There is, however, a special kind of light flowing through all things, which is for Bruno in some way divine light, for it illuminates everything from within, including the inquiring mind, and resides ultimately in the primary atoms. In his attempt to define what may be considered a lesser form of light, or the particular sources of light discernible in the universe, Bruno thinks of such light as glowing from multiple centers, or stars, but also from living things on the colder earths. This does not mean that such light should be seen as fractured or fragmented, for it is the element that unifies all single things. But its concentration changes and varies in a play of eternal vicissitude within the infinite whole. Bruno's view is thus fundamentally opposed to the Platonic one, which he repudiates in no uncertain terms. There is no transcendental light from above casting shadows on the wall of the cave; and even the neo-Platonic divinization of those shadows as emanating

13. See Bruno 1879–91, I, ii:301–18.

from the light of the divine is included by Bruno in the "Platonic fantasies" which he cannot share. The human mind can postulate but not see divine light; if it were to be seen, it would only blind the mind. Neither, though, can it see pure shade, which would equally amount to nonbeing and the collapse of the ordered whole. That order, and the reason of its intelligibility, is founded, for Bruno, on light traveling through a more or less subtle material medium. Together they form the infinite shadows of universal being. Anything that might lie beyond this infinite substance it is beyond the capacity of the human mind to know.

"The nature of composite things cannot be conceived of as separate from their composition any more than accidental things can be conceived of as separate from their subject." Having expressed his final rejection of Platonic and neo-Platonic transcendentalism, Bruno goes on to criticize once again the abstract universal objects of the logicians and the mathematicians. Here, however, his tone is more cautious. Conceding that such abstractions might be necessary for the construction of a physical science, Bruno is now ready to allow speculation on these lines provided that no illusion is nursed of their correspondence to the world of physical things. This way of formulating the problem seems to lead back to the traditional Aristotelian distinction between the potential and the actual and could hardly constitute Bruno's last word on the subject of scientific inquiry. In fact, the final chapter of the last book of his trilogy returns to the question of method in far more challenging and original terms.

Bruno here uses the snail as a symbolic form of the workings of nature, a body that gradually unfolds outward from its inner shell into a less compact and more fully expanded mass, ever ready to contract into itself again without ever reaching a final form of self-determination. The problem is how to penetrate the secret of the laws that govern this apparently random and irregular assumption by all beings of the various forms through which they express their nature. Bruno does not despair of being able to do this, for nature does obey laws. Her laws, however, are not to be understood universally, in the abstract; and for Bruno it remains clear that mathematics and logic can never say anything about the true workings of the natural world. At most they may become useful tools for computation and analysis. The true laws of nature can be understood only by the scrupulously attentive intellect observing the behavior of the bodies in space and time: "the splendour of the sun, the species of things that generate themselves from the entrails of our mother earth, explicating their true images in the appearances of the innumerable animated things which in the immense spaces of a single sky shine, live, feel, understand and applaud the one, best and maximum being."

In his final page, following this celebration of the infinite vicissitudes of infinite being, Bruno defines the attitude and activity of the new inquirer into its order and pattern. The metaphor he uses is a traditional one, which will nevertheless continue in the new science for many decades to come. The new scientist will be none other than the traditional piper, or faun, intent on penetrating the harmonies of nature. The search, however, is no longer to be confused with that of pastoral poetry, whose elegant, honeyed notes are associated with the servile Ganymede. The new faun is not going to serve Jupiter but himself: he needs to know how the harmonies of nature work so he can harness them to his needs. The process, however, will not be simple, for with the disappearance of a primitive "immersion" in the natural world, an alienation and distance have been created that render the process dangerous and uncertain. The new piper sounds a note which is not sweet but clear and well formulated. He waits for nature to respond: has his intuition been right? Is the note one nature will recognize? When the echo reaches back to him, he knows that something of the natural process has been understood. Now how will he proceed? Not like the loving Penelope, who waited patiently for the return of the wandering Ulysses. The new science has little time to wait; and Bruno admits that he may have tarried too long in the contemplative ardors, the heroic frenzies of a search for a total identification between subject and object. Now something quite different and more rigorously severe is at stake. When the piper hears the echo returned to him, he will no longer play its notes back again in an incessant chorus of praise. The age of Narcissus is over, and the decadent courts of the renaissance, with their eunuch voices, have had their day. The new piper, classless and pragmatic, will advance in the footprints left long ago by Pan (here we have one of Bruno's final homages to ancient Greek, pre-Socratic science) in an attempt to clutch and possess the order (the nymphs) he has just glimpsed in those fragmentary notes. Today some might be tempted to see this final moment as a rape; and Bruno explicitly uses the language of gender to close his work.

The metaphorical and mythical mode of expression used in this closing page may have deceived readers at the time, and still may do so, into thinking that Bruno is really talking about something other than what today we recognize as science. On the contrary, he foresees here the development not only of a scientific but of a properly technological and industrial society. Thus it is necessary to inquire a little closer into what conception of science Bruno is actually proposing. The question of method, as we have seen, was one in which Bruno made a less than clean break with the past, torn as he was between modern and ancient forms of mathematics and logic. Nevertheless, this final page of the *De immenso* does seem to develop his discourse on method in important ways that are interesting to consider in light of Ernan McMullin's essay "Conceptions of Science in the Scientific Revolution."[14] McMullin claims, in my opinion incorrectly, that "it was only with the blossoming of the 'new sciences' in the seventeenth century that for the first time philosophers, in constructing a theory of science, began to look hard at what was actually achievable in the way of a systematic knowledge of nature." It is my thesis, on the contrary, that McMullin's carefully argued treatment of his theme could well have included Bruno, who, in an earlier paper, he had excluded from his attention largely following Yates.[15] If we attempt to apply to Bruno the arguments developed by McMullin in his discussion of Descartes, Galileo, Kepler, Boyle, and Locke, we can trace in this page of Bruno's a conception of science that corresponds to the categories used in his paper.

McMullin derives from Charles Sanders Pierce three categories of early scientific investigation—deductivism, inductivism, and hypothetico-deductivism—which he uses to investigate the scientific methods proposed in the seventeenth century. Without attempting to force any of the figures he considers into the straitjacket of these categories, he tends to associate them with one in particular. Deductivism, seen as related to the Aristotelian forms of demonstration or the intuitive insight into the essential properties of nature, is related to the kinematics of Galileo. Inductivism, or the generalization of natural laws from the observation of singulars, is discussed with relation to Bacon. Hypothetico-deductivism is seen as the testing of a hypothesis by the observational consequences deduced from it and is discussed in relation to the method of Descartes. McMullin sees Newton as the figure in whom all these methods "collided" in a mode of deducing from the phenomena which declined both the search for causes and the feigning of hypotheses.

It appears to me that in this discourse on method proposed by McMullin, Bruno, in this final page of the *De immenso*, would be found somewhere between Baconian induction and the hypothetico-deductive method associated with Descartes. Much work on Bacon has revaluated his inductivism by distinguishing it from a radical form of empiricism which moves without any previous conceptual plan through the chaos of

14. McMullin 1990:27–92.

15. See McMullin 1987, and, for an interesting use of McMullin's essay, Badaloni, 1988:139–41.

single things. As Paolo Rossi pointed out, this kind of experience is seen by Bacon as "mere groping as of men in the dark, who feel all round them for the chance of finding their way, when they had much better wait for daylight, or light a candle before taking a step."[16] In an essay titled "Francis Bacon and Man's Two-Faced Kingdom," Antonio Pérez-Ramos has seen Bacon's driving force as "the vindication of a prototype or paradigm of knowledge and of a criterion to gauge it that he . . . nowhere appears to have fully articulated."[17] The Baconian paradigm needs collateral security which results from the production of effects; and Pérez-Ramos sees Bacon's method of obtaining this as open-ended, a procedure through *vindematio*, or different "vintages" of experimental monitoring, which covers increasingly large numbers of empirical phenomena with each step. This metaphor of the Baconian "two-faced kingdom" can clearly be related to Bruno's metaphor of the piper sending out his harmonic theme into the hills and dales and waiting to hear whether nature will return them to him with an echo.

In such formulations of Baconian induction, the difference from the hypothetico-deductive method of Descartes becomes one of emphasis rather than kind. For Bacon, the moments of increasingly complex inquiry into the single things of nature are the really "revolutionary" moments of the new scientific project: witness the fervent activity of the members of Salomon's House in the *New Atlantis*, crowned by the achievement of the sages who finally succeed in establishing the natural laws which the single phenomenon studied can be seen to obey. For Descartes the law of reason governing the founding concepts in the mind has precedence over the moment of empirical research: geometry thus becomes the foundation of the new science. We have seen Bruno finally, if almost reluctantly, concede this task to geometry too; although it is important to emphasize that what has been called the "grey ontology" of Descartes hardly corresponds to Bruno's natural world of constant and life-giving vicissitude.[18] It is nevertheless clear that the starting point of

16. Rossi 1984:245–60. Rossi is referring to a celebrated passage in Bacon's *Novum Organum*, I, 82, in *The Works of Francis Bacon* (London, 1887–92), 1:189 and 4:80–81.

17. Pérez-Ramos 1993:140–66.

18. There is, however, much discussion at present of the true nature of Descartes's philosophy of science. For a view that Descartes was really a practicing scientist who "somewhat unfortunately" wrote a few short and relatively unimportant philosophical essays, see Clarke 1982. For an alternative view of Descartes's sense of the confusion normally made by the mind between the mental and the material and the importance of his "Skeptical Therapy," see Garber 1986:81–116.

Bruno's inquiry is the tune played by the faun on his pipes, which, being a tune, must be founded on an order and harmony already present in the mind. That the starting point of the inquiry should be the light of reason had already been stated by Bruno in the opening pages of the *De triplici minimo*; just as in the *Regulae* of 1628, Descartes will find the starting point in "the undoubting conception of an unclouded and attentive mind which springs from the light of reason alone."

The second phase of the inquiry seems to me, in these final pages of the *De immenso*, to show in Bruno a confidence in success that was often absent in his previous works. Descartes posited in the sphere of matter a necessary rule of law determined by the infinite perfection of a transcendental God. Bruno had no such God to appeal to; but he nevertheless sees such laws as guaranteed by the association of matter with light. His nature is erudite, imbued with an eternal and arcane wisdom. Within it, a vital dialectic of contraries can be traced through the dialectical movement that dominates the mind. Thus both the movement of the mind and the natural world it perceives are more varied and dynamic in Bruno's philosophy than in Descartes's, which is why he was read by the post-Kantian idealists. Nature will respond to the piper's notes when they are true harmonies and will reveal to him the forms of its footprints. With Eddington in mind, we are immediately tempted to ask once again if the footprint is not our own; and indeed, Bruno's metaphor of the piper and his echo seems almost designed to highlight the danger of self-referentiality of which the modern philosophy of science is so aware. But this is not Bruno's concluding note. Rather, he closes his work by positing a third stage of synthesis in which the piper proceeds toward a moment of union with the natural world which will allow him to act and produce within it. Bearing in mind the threefold logic outlined in the pages on the *tectratys* in the *De monade*, this final phase must be seen as the moment of action and speech. Comforted by the echo returned to his own idea of order by the forms of nature, the new scientist begins to manipulate and organize his world.

In a chapter dedicated to Bruno in his monumental work *The Genesis of the Copernican World*, Hans Blumenberg has called attention, apparently with approval, to the critique developed by Hegel of this final moment of Bruno's thought.[19] Blumenberg points out that Hegel appreciated in Bruno's work "a great beginning of the effort to think unity" through the elimination of the Aristotelian dualism of matter and form and the thesis that matter has "life in itself" and is thus capable of endless trans-

19. See Blumenberg 1975/1987:353–85.

formations. Hegel, however, regards Bruno's effort to think unity—to present "the logical system of the inner artist," of the productive idea, in such a way that the configurations of external nature correspond to it—as unsuccessful. What emerged, he says, was only a universe that dissolved in the all-devouring mataphorics of the organic, which had evaded the exertion of the concept. The moments of the world process, writes Blumenburg quoting Hegel, are only "collected," "enumerated," or described and not developed in a stricter sense. In other words, the *coincidentia oppositorum* is not the dialectic required. The criticism is certainly cogent and is corroborated by Bruno's reference to the art of memory as his principal mental tool. We have seen, however, that above all in the final *De imaginum, signorum et compositione*, Bruno attempts to use methods of compression and association of signs to elaborate messages that are more than descriptive and that suggest relationships and connections not previously perceived. Furthermore, he explicitly insists in final page of the *De immenso* that his piper will not be so lazy as simply to repeat the tune returned to him by nature. Admittedly it is not yet clear what exactly he will do, outside the metaphor of the embrace. Perhaps the return to the logic of mathematics in the final *Praelectiones geometricae* indicates Bruno's awareness of a possible way to be taken, even if he himself has to be seen as standing rather cautiously at the temple door of mathematics rather than as a practitioner himself. Certainly his unrequited admiration of Tycho Brahe, as well as his more ambiguous enthusiasm for Fabrizio Mordente and his new compass, indicate Bruno's conviction that precise measurement was going to prove essential, while his repeatedly expressed conviction of the founding presence in the universe of entities impossible to see indicates his awareness of the necessity to develop new instruments for observation and measurement capable of reaching out both toward the further regions of outer space and toward the primary particles of matter. It is certainly true that in Bruno such developments of the scientific inquiry are present only as prophecy, still only vaguely understood and defined. Looking back on the historical process now, however, we can see that in the 1580s, when the name of Copernicus was only beginning to be mentioned surreptitously in the high places of European culture, it is remarkable that they were present at all.

Bruno's development of the Copernican theory, which in its original formulation attempted no more than a redefinition of the relative positions and movements of the neighboring planetary system, into a full-scale picture of a unified, infinite universe; his elaboration of a theory of matter that attempted to guarantee, throughout that universe, the idea of intelligible natural laws; his attempt, though still uncertain, to define an epistemology that would allow forms of practical, technical interaction between the mind and nature: these were the achievements that culminated in the final page of the Frankfurt trilogy with its conviction that a new era would enjoy a fuller knowledge and possession of the natural world.[20] Whether that possession would lead to new forms of alienation between nature and the mind, and whether such alienation would eventually prove to be subject to new and superior forms of reconciliation, is a subject about which Bruno was already meditating before the project had fully begun. It is for this reason (and not, as a nineteenth-century commentator such as Tocco believed, because of a childish attachment to external ornament and frills) that he refused to abandon the modes of discourse proper to myth. The scientific project, according to Bruno, would be valid only if it developed as a superior form of myth: a story told about the ways of the natural world which must remain close to the affections of the mind in order to be conversant with the deepest motives that determine the character and behavior of the infinite objects that populate an infinite world.

20. Pérez-Ramos, in my opinion correctly, asserts of Bruno that he "rejects the primacy of contemplation and argues that, where there is the power to make and to produce something, there is also the certainty of that something being known" (1993:145).

12 An Afterword: The Ethics of Scientific Discovery

One of the most brilliant and genial insights of Francis Bacon, at the beginning of the seventeenth century, was his presentation of the new scientific investigations into nature in the context of the Genesis myth of the Fall of Man. If the Fall had signified a forgetting of the knowledge of paradisiac natural surroundings, the new science would gradually rediscover this pristine state of natural knowledge and reconquer for humankind a place in sight of the abode of God: "For man by the Fall fell at the same time from his state of innocency and from his dominion over nature. Both of these losses, however, can even in this life be in some part repaired; the former by religion and faith, the latter by the arts and sciences."[1] With one fell swoop, Bacon thus cut through the Gordian knot of the conflict between science and Christianity which had been building up menacingly during the sixteenth century and would continue to agitate the Catholic part of Europe for many decades to come, finding its essential point of reference in the trial of Galileo. England, in contrast, by developing Bacon's insight, could proceed toward the establishment of a Royal Society devoted to the promotion of an empirical science which, as its founders were fond of emphasizing, contemplated no conflict between science and religion and placed itself at the service of both its church and king.[2]

1. See Francis Bacon, *The Advancement of Learning,* quoted in Prior 1968:140–63. For Bacon's use of the story of the Fall, see in particular 150–55.

2. The idea is clearly expressed in the introductory "Epistle to the King" in Sprat 1667/1959. Cowley's dedicatory poem written for this volume expresses the same idea by figuring Bacon as the Moses who had led England within sight of the promised land.

In this context, Bruno's Catholic background and formation are clearly in evidence. For him, at the end of the sixteenth century, Bacon's solution of the conflict would not have been possible. For Bruno agreed with the theologians that the new vistas of human knowledge opened up to the qualified layman by the new science clearly contradicted the traditional assumption that true knowledge of things, both natural and divine, should be limited to God and to his ministers on earth. Bruno's reaction to this contradiction, however, was very different from that of the theologians. For Bruno it was necessary to rewrite the traditional classical and biblical myths about the dangers of knowledge, presenting Prometheus or Icarus or Adam as positive rather than negative heroes and their stories as success stories rather than tragic failures inspired by presumption or pride. Those who refuse the challenge of pursuing new forms of knowledge, writes Bruno in the *Cabala of the Pegasean Horse*, are asses, or beasts who have closed up their hands, catching all five fingers under a single nail, "so that they remain unable to stretch out their hands like Adam to pluck the forbidden fruit from the tree of knowledge, and in consequence they remain without the fruits of the tree of life, or to hold out their hands like Prometheus (which is a metaphor of the same kind) to take the fire of Jove to light the lamp of rational ability."[3]

Nevertheless, Bruno's view of the matter did coincide with Bacon's on one essential point. He too thought of the new science as a major movement of historical and intellectual reform, which would necessarily impinge on the moral status and ethical principles of its practitioners, as well as on those of the society in which it developed. In fact, for Bruno, the development of a new science became *the* true process of reform in the sixteenth century: an intellectual movement that could and should be compared positively with what he saw as the more compromised and less innovative revolutions proposed by the Catholic Counter-Reformation, the Protestant Reformation, or even the neo-Platonic renaissance.

Bruno's reading of the scientific revolution in these terms seems to me to be the common theme underlying the last three of the six Italian dialogues he composed in London between 1583 and 1585. These six dialogues are made up of an initial trilogy starting with *The Ash Wednesday Supper*, a work that proposes a heliocentric post-Copernican universe extended to infinite dimensions and populated by an infinite number of worlds. The *Supper* is then followed by two dialogues that investigate the consequences of this infinite cosmological picture—radically reformed

3. Bruno 1958:878; my translation. For the positive rewriting of the Icarus myth, see *The Heroic Frenzy*, ibid., 999, and Gatti 1989:81–89.

with respect to the traditional Aristotelian-Ptolemaic universe—concerning its relationship with the first cause and infinite divine power: *On the Cause, Principle and One* and *On the Infinite Universe and Worlds*. The second trilogy is centered on ethical problems which introduce the theme of reform, or rather of multiple reforms, of customs, of power, of faith, of ideas; and these reforms are introduced in terms that appear at first sight somewhat wild and bizarre. In the first place, *The Expulsion of the Triumphant Beast* presents a universal reform of the heavens conceived in terms of the forty-eight signs of the zodiac, which are expelled by a Jupiter who has repented of his misdoings, to be replaced by the long-exiled virtues. In the second place, in *The Cabala of the Pegasean Horse*, a reform of the life of the spirit is treated, in clearly satirical terms, as abortive, given that in the end the winged horse suffers a metamorphosis into an ass: *The Mercurial Ass of the Nolan*, as the epilogue of this work is titled. Third, a long sonnet sequence in the Petrarchan tradition, *The Heroic Frenzy*, develops the terms of the neo-Platonic anxiety for an ascent of the soul toward an ineffable, divine truth.

This second trilogy is considerably more complex and varied than the first one and presents problems of interpretation which remain unsolved even after the constant critical attention dedicated to these works, and to the various proposals for reform which they contain, both in remote and recent times. Two critical cruxes remain to be resolved in order to arrive at a coherent reading of these three final Italian dialogues. First, in what ways are these three works, apparently so different one from the other, linked together? Second, what position does Bruno assume with respect to the proposals for reform which he puts forward? The two questions are closely connected. Once Bruno's attitude to the proposals for reform at the center of these three works is fully appreciated, it will be possible to delineate the relationship between these proposals and the post-Copernican cosmological reform that lies at the center of the initial trilogy.

All three dialogues of the first trilogy are dominated by a character at times called Theophilus, at other times Philoteus, who claims in *The Ash Wednesday Supper* to be as close to Bruno as he is to himself. Although Bruno's own position, as it is made clear in the course of the dialogue, cannot be identified completely, and at all times, with that expressed by Theophilus, he is nevertheless clearly a mouthpiece who guarantees the status of the new cosmology of *The Ash Wednesday Supper* as the cosmology proposed by Bruno himself. Perhaps this identification between the authorial voice and the principal message proposed in the first three dialogues has induced the critical tradition to consider also the moral reform proposed in the first work of the second trilogy, *The Expulsion of the Triumphant Beast,* as a reform proposed by Bruno himself. John Toland, in a well-known account of the *Expulsion* sent to Hohendorff at the beginning of the eighteenth century, and then in a series of manuscript notes in his copy of the rare text of the first edition of this dialogue, distinguished between two distinct moments in the narrative, as Rita Sturlese pointed out in her presentation of these marginal notes: "one which could be called negative, in which the beasts which have triumphed for so long due to the foolish credulity of human beings are expelled from the sky, the other positive, which witnesses the installation of forty-eight virtues in the places of vice."[4] Toland considers this second moment of the universal history narrated by Bruno to be the vehicle through which the Nolan philosopher presents his own personal proposal for reform, or a "complete system of natural law."

Another well-known interpretation of this text in terms of a universal reform to be identified with Bruno's moral and religious position is that of Frances Yates, developed in her book on Bruno of 1964, which singles out in particular the pages of the third dialogue composed of a long quotation from the Hermetic text titled *Asclepius*.[5] Here there is to be found a celebration of the magical and natural religion of the ancient Egyptians, which, according to Yates, is to be considered the principal theme of Bruno's dialogue as well. Furthermore, in Yates's view, this Egyptian religion appears clearly from these pages to correspond to the religion favored by Bruno himself. This identification of Bruno with the message proposed in these pages quoted from the *Asclepius* is repeated by Michele Ciliberto, though he criticizes as oversimple and too unilateral the Yatesian interpretation of this text. Ciliberto compares the Latin text of the *Asclepius* used by Bruno and his translation of it into Italian and claims to have found linguistic elements that justify the choice of these pages by Bruno as a means with which to criticize indirectly the Protestant Reformation: a message directed toward his contemporaries which, in Ciliberto's opinion, was far more important to Bruno than a simple celebration of the ancient Egyptians.[6] Nevertheless, although proposing a somewhat different interpretation of the passage from the *Asclepius* than that of Yates, Ciliberto remains faithful to the tradition insofar as he continues to consider the message contained in these pages as corresponding to the point of view of Bruno himself.

4. See Sturlese-Pagnoni 1986.
5. See Yates 1964:211–15.
6. See Ciliberto 1990:149–57. In his 1986 book, Ciliberto had already compared the lament in the Latin *Asclepius* to Bruno's Italian translation, claiming that the Italian text is slightly altered in many places to adapt it to Bruno's own society and times.

We need to ask, however, whether this text, together with the two following works of the trilogy of moral dialogues, requires a different reading technique from that necessary in the previous trilogy, where Theophilus is explicitly presented as a mouthpiece for Bruno himself. Bruno proposes such a difference in the most appropriate place: that is, the "Explanatory Letter" of the first work of the second trilogy, *The Expulsion of the Triumphant Beast*. Here Bruno proffers to his reader a series of warnings which indicate that he is now clearly taking his distance from the moral message, as well as from the program of reform, which his text is about to develop:

> Now, in order that my intention in the present discourse should be understood by anyone desirous and able to do so, I protest and declare that, as far as I am concerned, I approve that which is commonly held to be worthy of approval by those that are knowledgeable and worthy, and disapprove of the contrary as they do. For which reason, I pray and implore you all that nobody should be so gross in their thoughts or so malign in their spirit as to suggest, to themselves or to others, that what is written in this volume is said by me assertively.[7]

This passage contains a note of irony, which may be a defensive precaution against possible criticisms or accusations. Nevertheless, it is clearly also claiming seriously for its author a measured distance from the subject he is treating. In this sense it should be related to the figure to whom the explanatory letter is explicitly addressed: Sir Philip Sidney, already praised by Bruno in *The Ash Wednesday Supper* as the most cultivated and noteworthy courtier of Elizabethan England.

As well as the decidedly Protestant religious position professed by Sidney, who in these same years will reveal himself as one of the most bitter opponents of the proposed marriage between the queen of England and the French duke of Alençon, a reference needs to be made to Sidney's pastoral work, *Arcadia*, of which he was writing the final version during the years of Bruno's stay in England.[8] Sidney's is a neoclassical Arcadia but

7. Bruno 1958:553; my translation.

8. Sidney 1593/1977. All references are to the final version of the *Arcadia* published by his sister after his death. The original version of the text, usually known as the *Old Arcadia*, was probably begun in 1577 and finished at the Countess of Pembroke's country house at Wilton in 1580. Sidney was working on his revision of the text, which remained incomplete at his death, during Bruno's stay in England. Although an edition of his partly revised text was published under the supervision of Fulke Greville in 1590, it is usually the fully revised *Arcadia* of 1593, completed and published by Sidney's sister, which is taken as the established text, although the *Old Arcadia*, rediscovered and published only

not a very utopian one. On the contrary, it is full of evil and subtle traps: a world in which two young princes have to learn how to move with Machiavellian caution but at the same time with a private and public moral steadfastness which Sidney stresses with severely didactic notes that reveal his Protestant-Christian inspiration. Nevertheless, accepting with elegance the ideal of "sprezzatura" developed by Castiglione in the text of his *Courtier*, which many consider the model on which Sidney constructed his public and intellectual image, he maintains a decorous distance from his own literary work. Already in the title, the book is not presented as Sidney's own Arcadia but as that of the Countess of Pembroke, the beloved sister for whom the work was written as a graceful act of homage.[9] Furthermore, the importance of "sprezzatura," or an aristocratic indifference toward one's own work (it should be remembered that Sidney never published anything during his lifetime, even if his works circulated in manuscript among the cultured English of the time), is once more underlined at the end of another text of Sidney's, his well-known *Defence of Poesie*, probably completed shortly before Bruno's arrival in England.[10] There, after passionately defending the values of art in general, and of poetry in particular, against the criticisms of the most radical Protestants, who wished to limit the use of language to didactic and religious ends, Sidney, in a brilliant final paragraph, challenges his reader not to take too seriously an activity which, seen from another point of view, is to be considered little more than a refined pastime. So it is hardly to be wondered at if, after this surprise farewell from his reader, Sidney, called on by Elizabeth to fulfill a very different task, leaves without hesitation for the war in the Low Countries against the Catholic forces of Spain which will lead to his premature and tragic death.

Bruno's warning to his reader in his explanatory letter not to take assertively what is said in the text thus appears to be modeled clearly enough on this attitude of "sprezzatura" assumed by its noble recipient. Nor should it be forgotten (as often happens in Bruno criticism) that the

in 1926, has found a few enthusiastic supporters among modern critics. For an interesting if debatable preference for, and interpretation of, the *Old Arcadia*, underlining its specifically Protestant ethic, see Marenco 1968.

9. A letter from Sidney to his sister, published in the text edited by Greville in 1590, describes the composition of the work: "Your dear self can best witness the manner, being done in loose sheets of paper, most of it in your presence; the rest by sheets sent unto you as fast as they were done."

10. See Sidney [1580?]/1965:141–42. The text was first published in London in two separate editions, several years after Sidney's death, in 1595. It appeared from the printer Olney with the title *An Apologie for Poetrie* and from the printer Ponsonby with the title *The Defence of Poesie*.

first Italian work published by Bruno, which appeared in France in 1582, was a drama—the *Candelaio*, based on the various and lively vicissitudes of downtown life in contemporary Naples—given that he invites his reader, again in the explanatory letter of the *Expulsion*, to remember "that these are dialogues, with speakers who express their own opinions, and who talk about the opinions of many others who also express their own point of view, each of them arguing their case with that zeal and eagerness which is appropriate to them."[11] In this case, the words of Jupiter in the moment he puts into effect his substantial and systematic moral reform are not to be taken as carved in the heart of the philosopher from Nola, but rather as the words which he puts in the mouth of a Jupiter not to be considered, as Bruno warns once again, "as too legitimate or valid a representative or lieutenant of the first principle and universal cause."[12] And clearly the warning is important with respect to the epic doings of a penitent Jupiter, who, helped by a satirical and cynical but, at the end of the day, always obedient Momus, assumes as his own the entire panorama of ethical and legislative values of the Graeco-Roman world, beginning with truth, prudence, law, and justice.

Who, then, is this Jupiter who calls the pagan gods to a council to propose to them a universal reform? Once again in the explanatory letter, Bruno promises to reveal the true sense of his fable only at the end, when the reader has "penetrated into the marrow of it." In fact, following the Sidneian taste for a surprise finish, Bruno too shows his hand only on the final page of his work, where, tired but satisfied after his long effort, Jupiter announces his intention of taking a supper of fish (an antique symbol for Christ, as numerous places in the Roman catacombs demonstrate) dressed with, among other things, a Roman sauce.[13] This is clearly a reference to the Catholic doctrine of the mystery of the Eucharist as a transubstantiation of the bread and wine into the body and blood of Christ: a discussion of the philosophical implications of the Mass being, as Alfonso Ingegno has underlined in a book of 1987—which has the merit of also stressing the importance of Bruno's reading of the works of Calvin—a central element in the economy of this work.[14] And yet even Ingegno seems not to have realized Bruno's exact intention. Just as Sidney offers his sister a classical-pastoral work outlining a program of reform in which the moral values of the ancient world unite with the values of Protestantism, so Bruno offers him a classical-cosmological work that paints a

11. See Bruno 1958:553–54.
12. Ibid., 555.
13. "accomodato con salza romana": ibid., 829.
14. See Ingegno 1987.

picture of that other reform, produced in the other part of Europe through the Council of Trent. Better not to forget its dimensions, the power behind its systematic thought, its pretensions to universality. For Bruno, after many years as a Dominican monk, followed by a dramatic journey in exile through a Europe divided by fierce wars of religion, could have had few illusions as to the epic dimensions of the Council of Trent, which Paolo Sarpi, when he started to narrate the same story some years later, would call "the Iliad of our century." Perhaps it is no coincidence if the text of Sarpi, who may have met and spoken with Bruno after his return to Venice in 1591, was sent in secret to London, where it appeared for the first time published by the printer John Bill in 1619.[15]

The interpretation of the *Expulsion* as an antipapal satire is not new. It was proposed by Leibniz in a letter of 1708: "Je crois d'avoir vu ce livre autrefois, et qu'il est proprement contre le Pape."[16] Leibniz, who is commenting on the crusade in favor of Bruno's works undertaken by Toland in the freethinking circles of the Enlightenment in the Europe of the early eighteenth century, refuses to go further into the subject, preferring to recognize in Bruno above all the proposal of an infinite universe. Perhaps it was the caution shown by Leibniz which led to the general acceptance of the alternative thesis put forward by Toland, who identifies Bruno in person with the reform of the skies proposed in the *Expulsion*: a reading of the text which continues to prevail among a generation of Bruno scholars still deeply concerned with the radical pantheism of Toland. Yet it may have been Leibniz who came closest to understanding Bruno's text, accepting the interpretative key offered by Bruno in his explanatory letter. For if the text is read as a satirical account of the implementing of the Counter-Reformation, many of the principal themes, still at the center of critical attention, appear in quite a different light. For example, that continuous anti-Protestant attack, seen by Ciliberto as the dominating theme of Bruno's thought, appears quite differently when it is considered as the dominating theme of that Catholic world which was founding its impressive *renovatio* precisely in terms of a repudiation of the justification by faith alone, of a devaluation of works, and of the enslavement of the will to divine predestination. These are anti-Lutheran polemics which Bruno must have known by heart from infancy, and which appear in the *Expulsion* with a rhetorical and repetitive insistence that tends to belittle them and render them ridiculous. Indeed, how could it have been otherwise in a text offered publicly as a gift to Sidney?

15. See the introduction by Corrado Vivanti to Sarpi 1619/1974.
16. See Ricci 1990:265.

If one asks, instead, why Bruno chose to identify the papal figure with Jupiter, it is useful to refer to one of the most erudite pages of Frances Yates.[17] In a discussion of Pico della Mirandola and cabalistic magic, in the introductory part of her book on Bruno, Yates refers to a lecture given by Fritz Saxl in November 1945 at the Courtauld Institute in London in which he noted the strangeness of the frescoes painted by Pinturicchio for Pope Alexander VI in the Borgia apartments of the Vatican.[18] Here, with a reference to the family crest of the Borgias, which was a bull, Pinturicchio represents the pope as the Egyptian bull, Apis, surrounded by Egyptian religious symbols including an ornate pyramid. Pinturicchio is referring, in Yates's view, to the acceptance of Pico's cabalistic Hermeticism by Alexander VI after the severe disapproval of the preceding pope, Innocent VIII. To underline the sense of this acceptance, Pinturicchio, in this same series of frescoes, represents the goddess Io, who fled into Egypt, where she took the name of Isis, seated on a throne next to two masculine figures identified by Saxl and Yates as Moses and Hermes Trismegistus. A pictorial continuity is thus established between the Egyptian religion, the Hebrew, and the Roman Catholic, represented in the figure of the papal bull, which is also Jupiter and therefore the sun. Such a continuity is part of a particular climate of humanistic thought, far removed from the historical vision of Protestantism which postulated an incommensurable divide between pre-Christian and post-Christian history.

It is a pity that Yates forgot about this page of hers when she came to discuss the Egyptian section of the *Expulsion*, which she considers a pure celebration of the Hermetic religion on the part of Bruno. If she had paid attention, as Bruno advises in the explanatory letter, to the context in which each speech occurs in the development of the work, she would have realized that it is Isis and not the voice of the author who pronounces that celebration.[19] To be precise, Isis is called on by Jupiter to solve the problem of Capricorn, which, like the other signs of the zodiac, is about to descend from its place in the heavens to be replaced by its appropriate virtue. At that point Isis pronounces a passionate defense of the venerable beast, pointing out that the Egyptian religion adored the divinity in all things, discovering it even in crocodiles, cocks, onions, and turnips. This does not mean that they should be considered idolators, maintains Isis, as the Egyptians knew how to refer every recipient of the

17. Yates 1964:115–16.

18. Saxl's text was published in two parts in *Lectures* (London, 1957), with the title "The Appartamento Borgia," 1:174–88, 2:115–24.

19. For the whole episode regarding Isis's defense of Capricorn, see Bruno 1958: 775–97.

divinity to a spirit of spirits, or fountain of ideas above the sphere of nature. The Hebrew cabalists held a similar vision, Isis admits, but they are to be considered as simple imitators of the Egyptian culture that held them so long in captivity. To sum up, Isis wants the religion of those whom she calls "my Egyptians"—that is, the religion of which she herself represents one of the principal divinities—to be the truly enlightened religion, stimulating the sarcastic comment from Momus that all historical religions have claimed with certainty to be in the light and their rivals in the shade.

It seems clear that what is going on here is something much more complex than a simple celebration of the Hermetic religion of the Egyptians. Nor is it to be considered particularly anti-Protestant, in my opinion, except to the extent to which all the historical religions become involved in a radical process of relativization which, in any case, was the aspect of the work most appreciated by Toland. When Sophia, who is narrating the progress of the Council of the Gods to Saulino—a skeptical layman who nevertheless follows with attention the development of Jupiter's universal reform—proposes the long quotation from the *Asclepius* to illustrate the point of view of Isis, the fact is certainly not without importance but should nevertheless not be overvalued. For later on Sophia, together with Saulino, develops a lively criticism of Jupiter's decision to accept the defense of Capricorn pronounced by Isis and to leave it in the sky undisturbed. Saulino and Sophia, for their part, want to be free of the beast, which they find too aggressively materialistic and not sufficiently spiritual. Its horns, which seem to them like crowns decorating the heads of terrestrial and ecclesiastic princes, are derided in terms (and here it is necessary to recognize an important contribution by Ciliberto) which recall above all the Erasmus of *The Praise of Folly*.[20] One may even wonder if Sidney, as one of the most distinguished members of Elizabeth's undoubtedly luxurious court, would have been able to accept such a page. In the years of Bruno's stay in England, however, Sidney's relationship with

20. Ciliberto in both his books on Bruno (1986 and 1990) insists on the importance of Erasmus's influence to the extent of calling him the "maestro" of Bruno, whose trouble at the Monastery of San Domenico in Naples started when he was caught reading the *Opus epistolarum* of St. Jerome with a comment by Erasmus, whose works were already on the Index of forbidden books. See Canone, ed., 1992:70–75, for details of the Neapolitan proceedings against Bruno. Ciliberto considers the parabola of Erasmus's influence on Bruno to have exhausted itself in the pages of the *Cabala of the Pegasean Horse*, where Bruno criticizes what he sees as the distorted foundations of Christian culture, symbolized by the figure of Paul of Tarsus: "Erasmo ne fa l'apologia nelle pagine finali dell'*Elogio*; Bruno lo presenta come straordinario archetipo dell'asinità cristiana." See Ciliberto 1990:167.

the queen was far from cordial, and he was often away from the court. Furthermore, the ecclesiastical aspect of Sophia's and Saulino's criticisms of Capricorn are perfectly in line with the Protestant condemnation of sumptuous robes and ceremonies with respect to which the reformed religion was proposing a more severe and humble idea of priesthood. The work thus remains, even in these much-discussed pages, perfectly coherent as a gift carefully prepared for the Protestant Sidney.

Although this reading of the *Expulsion* stresses the necessity of considering it in the light of its connection with Sidney, it is not my intention to enroll Bruno in the ranks of Protestant Christianity, as happened in certain German circles at the beginning of the eighteenth century, well described by Saverio Ricci.[21] The sense of these three final Italian works, in my opinion, is rather to be found in a transition from an intellectual sphere dominated by a vision of the world in essentially theological terms to an intellectual sphere dominated by a vision of the world in essentially philosophical terms. In this passage from theology to philosophy all forms of revealed religion receive harsh treatment, but above all the Christian religion that dominated the life and culture of the Europe of the sixteenth century, often through violence and oppression. Bruno directs this criticism, in my opinion, with equal severity at both forms of Christianity, the Roman and the Protestant, often joined together in the same pages in these three works. Thus the ass, which is the central image of the second of these three Brunian works, composed of two parts, *The Cabala of the Pegasean Horse* and *The Mercurial Ass of the Nolan*, is called on by Jupiter to occupy two distinct places left vacant in the heavens, thereby completing the progress of the great reforming council of the Roman world of the sixteenth century; while the entire context in which this work develops recalls Bruno's own youth in his monastery, beginning with that undistinguished Don Sapantino to whom it is ironically dedicated.[22] Nevertheless, the increasing importance of certain themes such as a new cabalistic and mystical spirituality, or an intellectual sloth based on lack of faith in the powers of the human mind, can certainly be applied equally well, if not better, to the Protestant sphere of the Christian world, in which Bruno was situated and at which he could therefore aim his criticisms only in indirect and allusive ways. The intellectual model toward which Bruno directs his reader in these two brief pamphlets is what he

21. The conviction that Bruno should properly be seen as a martyr for the cause of Protestantism came from Christoph August Heumann. See Ricci 1990:362–76.

22. Spampanato 1988:61 identified an undistinguished Don Sapatino Savolino known to Bruno in his youth, and presumed that he was ironically raised to the rank of abbott and bishop by Bruno himself.

calls the "Pythagorean academy" whose door, once again with the taste for a surprise ending, appears only in the final pages of the book, clearly opposed to the theological heaven generally, and to the Christian in particular, without subtle differences or distinctions.

It remains to be clarified in what terms Jupiter calls specifically on the ass to cover those two places left vacant during his universal reform: one of them the place of the Great Bear, in the northern hemisphere, which is covered by Asininity "in the abstract," the other of the River Eridano in the southern hemisphere, which is covered by Asininity "in the concrete," so that the entire theological heaven of contemporary Europe appears, culturally, under the dominion of the ass, defined by Bruno in the explanatory letter as "the triumphant beast alive." In commenting on this text, Nicola Badaloni and Nuccio Ordine have underlined Bruno's double attitude toward the ass: one satirical, as a symbol of obstinate ignorance, the other a positive appreciation of the character of mild docility, of a creature obedient to the divine will: characteristics which, in a Christian context, determined the choice of the ass as the beast on whose back the Savior decided to make his entrance into Jerusalem.[23] I believe that this second attitude, although continually present in this text, is not so much to be referred to Bruno himself as to those who, historically, have praised the ass in these terms, and who for just that reason are subjected to strong ciriticism on the part of the same Saulino who, in the *Expulsion*, listened to the story of the council narrated by Sophia while now, in the new work, he conducts an active dialogue with the spirited layman Sebasto.

The three intellectual movements that have prized the symbol of the ass, which Saulino prepares himself to investigate and repudiate, are defined clearly already in the first dialogue. They are, first, the cabalists and certain mystic theologians; second, the Pirronians, Ephectics, and such like; and third, the Christian theologians beginning with Paul of Tarsus. Of these three intellectual traditions, Saulino proposes to consider the members of the first and third, the Hebrew cabalists and the Christian theologians, as substantially the same: they are those who "referred to an incomprehensible and ineffable principle" and for that reason "closed their eyes, exiled all proper attention and study, disapproved of all human thought, repudiated all natural sentiments."[24] This lack of faith in human thought is illustrated, above all with respect to the intellectual foun-

23. See Ordine 1987/1996 and Badaloni's introduction to Bruno 1994a.
24. See Bruno 1958:877–78.

dations of the Christian culture of the period, with the comic episode that occupies the first two parts of the second dialogue in which a character called Onorio remembers to have been, in some preceding incarnation, Aristotle himself.[25]

I am unable to agree with Nicola Badaloni in his claim that these pages are a pure celebration of the doctrine of metempsychosis, which seems to me also to be subjected to a treatment at least with a smile.[26] Furthermore, such an attitude would in no way contradict the Pythagorean academy proposed at the end of the dialogue but would remain coherent with the position already announced by Bruno at the end of his first printed work, the *De umbris idearum*, where he wrote that no philosophical school is to be either completely accepted or completely rejected, but all of them considered with a selective and critical stance.[27] And it is precisely this lack of a spirit of criticism with which Onorio is reproached, given that he admits no distance between himself and the Aristotelian philosophy, but insists on identifying himself with it completely and without criticism. What "the Philosopher" wrote becomes dogma; and such a way of reading him does not exalt him at all, but rather reduces the Stagyrite himself to the level of that asininity which Onorio accepts willingly for himself. Nor do the skeptical philosophers emerge from Bruno's satire in more positive terms. Rather, they too are taken to account by Sebasto, in the third part of the second dialogue, with an attack almost certainly directed toward the more extreme areas of Protestant Christianity. The radical Protestants, as Richard Popkin has demonstrated in his by now classic *History of Scepticism from Erasmus to Spinoza*, considered the impossibility for the human mind to arrive at the truth with its own powers an eloquent confirmation of the fundamental necessity for a faith in God.[28] The academic skeptics, in fact, are presented by Bruno's Sebasto as those who are ambitious "to form a new sect"; and they can do it easily precisely because by claiming the impossibility of the human mind arriving at the truth, they escape from "the fatigue of discovering how things are."[29]

This brief pamphlet titled *The Cabala of the Pegasean Horse*, then, not only completes Bruno's satirical representation of those theological reforms of the European heavens which had absorbed such a large part of

25. Ibid., 892–901.
26. See the introduction to Bruno 1994a:xxx.
27. See the final speech of Hermes in Bruno 1991a.
28. See Popkin 1979.
29. See Bruno 1958:901–10.

the intellectual energies of the sixteenth century, but at the same time subjects them to a lively criticism, leading the reader toward the door of the Pythagorean academy which he proposes in their stead. This choice of Pythagoras as a pivotal figure is to be considered as far more than a vague reminder of that *prisca theologia* which the studies of Paul Oscar Kristeller, Eugenio Garin, Frances Yates, Cesare Vasoli, and Brian Copenhaver have shown to be of fundamental importance for such a large part of the thought of the European renaissance. For Pythagoras had not only already been closely linked by Bruno, in the cosmological dialogue *The Ash Wednesday Supper*, to the name of Copernicus and to his newly heliocentric cosmology, but it furthermore comes to symbolize, most particularly in the context of these moral dialogues, the recognized historical founder of the discipline of philosophy: the inventor of the word itself—Philosophy—friend of knowledge. And it is clear that the idea of a dialogue, of discussion, and of collaboration even among differing points of view which animated the renaissance academies is of crucial importance here, becoming the principal theme of the brief epilogue of the *Cabala*, that is, *The Mercurial Ass of the Nolan*.[30]

At this point the condition in which a new era would hopefully develop is clearly seen as the substitution of a philosophical heaven, based on the idea of concord achieved through discussion, for a theological heaven dominated (as it then undoubtedly was) by the thunders of Jupiter. Bruno saw at once that this substitution would pose the problem of how to respond to the request of theology to enter the new academy. Would it be desirable or even possible for such an academy to close itself into a group of the initiated, or should it not open its doors also to approaches substantially different from its own? The brief pamphlet that ends this phase of Bruno's work offers a positive reply to this problem, without attempting to hide the difficulties. The dialogue is not going to be easy; there will be hesitations on either side. Furthermore, the mediation of some Mercury (to be read as a figure of ecclesiastical or academic influence, or perhaps of both) who insists on the admission of theology to the philosophical discussion is clearly a dangerous element. Nevertheless, the character called Micco Pitagorico, that is, the friend of philosophy, who guards the door of the new academy, knows that he has to admit and listen to the voices of all those who, with serious motives, wish to enter. To deny their entrance, as the Mercurial Ass itself points out in no uncertain terms, would be to destroy "the principles, foundations and body

30. Ibid., 913–23.

itself of your philosophy," precisely because of its claim to be based on the idea of free thought. Such is the sense of the brief, almost lapidary phrase pronounced at the end of the dialogue by Micco Pitagorico himself: "We are not deaf."

At this point Bruno can once again pick up his dialogue with Sidney: never really absent from these moral dialogues if it is remembered, as Andrew Weiner pointed out some years ago, that the figure of the bear was an important element of the Sidney family crest.[31] Thus the fact that asininity takes the place of the Great Bear in the theological heaven of the *Expulsion of the Triumphant Beast* has a precise meaning in the message that Bruno is directing to Sidney as the major representative of Elizabethan culture. This relevance is underlined by Bruno himself when the Great Bear is told to remain in the company, in the inferior world, of the Little Bear, which has already been sent to make friends with "the bears of England, or . . . the Orsini or Cesarini of Rome." It is not clear how far this destination of the Little Bear was intended as an offense to Sidney, who, in any case, appears not to have identified himself with the bear on his crest. On the contrary, the bear on the Sidney crest is shown face-to-face with the noble lion and seems to have a value of evil, corresponding to the way it is treated by Sidney himself in his *Arcadia*, where the innocent young girl, Pamela, is nearly killed by a "foul, horrible bear," only to be saved at the last moment by the shepherd Dorus.[32] It is interesting to note how Sidney in this page (with an evident reference to his family crest) compares the evil bear explicitly to the noble lion, underlining how the evil bear had attacked the innocent girl in the very moment in which the lion was absent: "fearing belike to deal while the lion was present." The bear thus seems similar to the ugly beasts kept by Cecropia, the attractive but evil enchantress who refuses to believe in a transcendent God. Nevertheless, its substitution in the sky by the cabalistic ass fails to represent for Bruno, as it failed to represent for Sidney in the *Defence of Poesie*, a true moral or cultural reform, which was not to be achieved by stifling the sciences or art. For this reason, the pages of "The Argument of the Nolan about the Heroic Frenzy written for the very illustrious Sir Philip Sidney" are addressed with evident respect to the author of the sonnets of Astrophil to Stella. Here Bruno is doing homage to the major representative of English neo-Petrarchism, which passed through the neo-Platonism of Ficino in an attempt to soften and humanize that rigid

31. Weiner, 1980, reads this passage as clearly offensive.
32. Sidney 1593/1997:179.

Protestant Calvinism to which Sidney, probable translator of the *Woorke concerning the trewnesse of the Christian Religion* by the Huguenot Du Plessis Mornay, remained faithful.[33]

That such an attempt to reconcile doctrines so different would yield contradictory and at times exasperated results has been well underlined by G. F. Waller in an essay titled "'This Matching of Contraries': Bruno, Calvin and the Sidney Circle," which considers Bruno the major representative of that Hermeticizing neo-Platonism so eloquently celebrated by Frances Yates.[34] It is undoubtedly true that the argument of the *Heroic Frenzy*, where the sonnet is shorn of all that superstructure of courtly love which it had acquired in order to bring it back to the metaphysical tension of its original Petrarchan model, could be defined as a learned and passionate *excursus* in Florentine neo-Platonism. The plot of the book, for example, could be summed up using the words written by Cesare Vasoli about the *De christiana religione* of Marsilio Ficino in a paper titled "Philosophy and Religion in Renaissance Culture": "The ascent to God is achieved by mounting the difficult 'ladder' of the virtues which frees the soul, true miracle of creation, so it can exert its dominion over the body and so rise from the sphere of the mundane to the supreme fount of Being."[35] And it is with notable stylistic subtlety and effort of thought that Bruno, calling to his aid other poets from the biblical Solomon to the co-Neapolitan Tansillo, traces the metaphysical impetus and the existential anxieties of the frenzied hero until, like a new Acteon, he finally reaches that moment of beatitude in which he sees and identifies himself with the nude Diana in a vision of divine truth which will render him from then onward "the happiest captured and subjected prisoner that any might wish to envy."[36]

Nevertheless, even here, in my opinion, it would be mistaken to identify this happy moment of neo-Platonic beautitude with Bruno's own philosophy as it had been developing after the cosmological discourse of the first Italian dialogue, *The Ash Wednesday Supper*. For Bruno himself, in

33. On the "Argument" of the *Heroic Furies*, see Farley-Hills 1992. Mornay, one of the principal French Calvinists of his day, met Sidney during his visits to England in 1572, 1577, and 1580. He dedicated his work to Henry of Navarre. The English translation, which Mornay's wife claimed in her diary to be by Sir Philip Sidney, was published in London in 1587, a year after Sidney's death, with the title *A Woorke concerning the trewnesse of the Christian Religion, by Philip of Mornay, Lord of Plessie Marlie, Begunne to be translated into English by Sir Philip Sidney and finished by Arthur Golding.*

34. Waller 1972.

35. See Vasoli 1988:45; my translation.

36. See Bruno 1958:1125.

some pages of a clearly autobiographcial flavor at the end of the *Heroic Frenzy*, situates this moment of his inquiry just before the journey undertaken by the nine blind men from the mountain of Circe, near Rome, where the enchantress has sprinkled a liquid on their eyes making them prisoners of an intellectual darkness: the darkness of a fascinating but material magic, unrelieved by intellectual endeavor. The resolution of the plot takes place on the banks of the river Thames where the chief nymph of England (certainly a covert reference to Queen Elizabeth I) opens another vial, also a gift from Circe brought by the blind men with them from Italy. This new liquid proves to be a balm which induces a process of sudden and exalted illumination as the nine exhausted travelers after truth gaze into the nymph's two shining eyes.[37] It is my opinion that this new illumination on the banks of the Thames is to be interpreted as a liberation both from the vision of Diana, an enchanting but perhaps illusory moon goddess who represents the metaphysical and mythical unity of nature in all her beauty, before whose gaze men become powerless prisoners, and from the baser enchantress Circe, who is described in the very last words of the book as a source of "dark blindness, desperate thoughts, and bitter pains." Which is to say that the search for a unifying metaphysical principle in nature as well as the praxis of a naturalistic magic, both inherent in Ficinian neo-Platonism, are repudiated by Bruno in this autobiographical allegory of his philosophical search. For although they are recognized as moments of an undoubtedly dynamic process of thought with respect to the theological heaven of the various forms of Christianity satirized in the previous dialogues, nevertheless they appear as ultimately inadequate as paths of entrance into a profoundly renewed world.[38] That world will open up before the nine blind men in search of new forms of truth only in the moment when they see the two suns represented by the eyes of the nymph of the river Thames: one sun "previously lost but now recovered," which can be identified with that of Pythagoras, and another sun "newly revealed," which can be identified with that of Copernicus. It is, then, with Bruno's new post-Copernican cosmology, based precisely on the idea of a plurality of suns, that the image of the ultimate good on earth is revealed; and the second trilogy of Bruno's Italian dialogues leads, with a process of perfectly clear and coherent circularity, toward the opening of the first work of the first trilogy:

37. Ibid., 1140–73.

38. Ciliberto has commented on the particular interest acquired by Acteon's hunt for the image of Diana in the comment by Octavio Paz in an essay composed between 1972 and 1976 on the art of Marcel Duchamp: see Ciliberto 1994:315–21.

that *Ash Wednesday Supper*, in which the infinite cosmology of the Nolan is for the first time revealed.

The last pages of the *Heroic Frenzy* represent, then, in my opinion, the moment in which Bruno moves beyond the Hermetic neo-Platonism which is the subject of the work itself: beyond the search for a contemplative vision of ecstatic illumination in which the subject dissolves itself in its intellectual object. With the beginning of the cosmological discourse in *The Ash Wednesday Supper*, which is presented by Bruno as a further development with respect to the unhappy experience at Oxford, Bruno makes a clear break with his own intellectual past. It is precisely the story of that intellectual past, and of the mode of arrival at that moment of rupture, which is narrated by Bruno, in my opinion, in the final trilogy of Italian dialogues. It is almost as if Bruno, having defined in the first trilogy the cosmological picture within which, from then on, his philosophy will develop, felt the necessity to clarify with himself and with his readers the stages through which he, and the century in which he lived, had arrived at that new point of departure, for Bruno truly revolutionary. What was at stake was the possibility of living in a new world, with a new physical structure that opened up, above all with the extension of the Copernican heliocentric universe to infinite dimensions proposed by Bruno in the *Supper*, new and different metaphysical, historical, and moral prospects.

This sense that everything that had been thought by his century outside the Copernican revolution represented, in some way, an old and decaying world is perhaps the idea which lies at the basis and informs the second trilogy of Italian works. It is announced with great clarity right at the beginning of the trilogy in the explanatory letter to *The Expulsion of the Triumphant Beast*. "This world," writes Bruno (and that is the world of the ecclesiastical prince of his time), "based on the fantasies of foolish mathematicians, and accepted by no less wise physicists, among which the Peripatetics are the most vain, not without present gain: first of all divided as if into so many spheres, and then signified by forty-eight images (according to what they understand as the primary division of the eighth sphere of the stars, commonly called the firmament), becomes the starting point and subject of our work."[39] And it remains the principal subject of the work up to the end of the *Heroic Frenzy*, given that even the more dynamic and subtle philosophical inquiry of Hermetic neo-Platonism continues to develop within this closed, Aristotelian-Ptolemaic universe, depriving it ultimately, for Bruno, of the possibility of reaching truly fruit-

39. Bruno 1958:560.

ful results. In the end Sidney himself becomes, in this sense, involved in the criticism directed by Bruno toward that old, closed world: Sidney the soneteer, who had himself "hunted" his prey in neo-Platonic terms, like the Acteon of the *Furori*. So it is no coincidence if it is in the first place to Sidney—if one accepts the identification common to the critical tradition that the anonymous knight seated at the head of the table during the Ash Wednesday supper was Sidney himself—that the cosmological discourse of the Nolan will be delivered. It is Bruno's introductory lecture to aristocratic London, not only on his new post-Copernican cosmology but also on a radically new, more rational form of inquiry into natural things.

The reading of the final trilogy of Bruno's Italian dialogues that I have just developed thus considers them not as an expression of his own mature philosophical position but as an inquiry into the historical and intellectual background that lies behind the new philosophy he had proposed in the preceding trilogy. In this sense, as Bruno underlines, the three moral dialogues are not to be taken "assertively" but as warnings not to succumb to those intellectual temptations that Bruno himself had responded to, with considerable intensity, in his younger years. They claim to have a primarily therapeutic value in preparation for the discovery and the full understanding of a truly new philosophy. That philosophy had already been presented in *The Ash Wednesday Supper* as the birth of a careful inquiry into nature to be undertaken outside of any revealed religion or traditional school of philosophical thought, but in the context of what Bruno considered, in the wake of the Copernican revolution, a new and truer cosmology. Nevertheless, the new science which is proposed is not seen as separated off from the proposals for moral, social, or intellectual reform which characterized his century and are considered at length in the second trilogy of moral dialogues. On the contrary, all that Bruno considers most just and valid in those proposals should accompany the birth of the new science; and insofar as such reforms are now proposed within a cosmology happily renewed and reformed, they will have a chance, in Bruno's opinion, of reaching positive results. For only in the context of an infinite universe, in which infinite worlds circle around their suns according to a centrifugal law that becomes a fundamental characteristic of matter, can there be a realization of that fruitful communication, without fractures or caesuras, between the divine mind, the infinite material substance, and an active human mind which is at the basis of Bruno's metaphysical and epistemological doctrine. So, in *The Ash Wednesday Supper*, Theophilus points out that his Pythagorean school not only proposes a universe which is once again heliocentric, extending it to infinite dimensions, but it further expects that within such a universe there will be a return of the "lengthy lives" of past ages, the "noble inventions," the "correct prophecies," the "inviolable sacraments," the "familiarity of good and protective spirits."[40]

The outcome is thus an identification between ontology and ethics, a link between moral and epistemological inquiry, which is not a part of the modern consciousness. Today, we expect the discoveries made by modern science to produce new therapies, new practical advantages; but we do not contemplate the possibility that they may improve our political structures, our religious practices, or our social conventions. Bruno hoped for precisely that, with an expectation that must be recognized, historically, as having been very soon disappointed. Already at the beginning of the seventeenth century, the clear distinction made by Francis Bacon between first and second causes represented a separation of spheres not contemplated by Bruno. It is true that the Bacon of the *New Atlantis* seems to have second thoughts on the subject, proposing a new figure of the scientist who is also a founding father of the society in which he lives: priest, prince, and counselor as well as being a scientist by profession. But the *New Atlantis* was already presented as a utopia; and in the same years of its publication the Cartesian dualism intervened to give another harsh blow to that unity of thought still present in Bruno. So that when, at the end of the century, Isaac Newton dedicated much of his free time from scientific research to the biblical and alchemical studies that today are at the center of so much attention, he hid them away in private papers, eliminating most references to them in the scientific publications that at the time assured his fame. And yet today, when philosophical inquiry is attempting to close the fractures opened at the beginnings of the modern world between subject and object, matter and spirit, Bruno's attempt to become the prophet of a new science in which dynamic thought moves to unite itself to the vicissitudes of an infinite universe that absorbs into its forms all the powers and the perfect order of God may appear to us as a historically defeated alternative, but not without interest. The idea of moral and spiritual as well as intellectual reform becomes identified with a methodical inquiry into the natural world because every deepening of understanding of the order inherent in the infinite substance will be also, if not above all, a discovery of the sacred well, or a pursuit of infinite truth.

40. Ibid., 44.

Bibliography

Abbott, G. 1604. *The Reasons Which Doctour Hill Hath Brought, for the Upholding of Papistry, Which Is Falselie Termed the Catholic Religion*. Oxford.

Abraham, G. A. 1983. "Misunderstanding the Merton Thesis: A Boundary Dispute between History and Sociology." *Isis* 74.

Alhazen [Ibn Al-Haytham, 11th cent.]. 1989. *Optics*. Ed. I. Sabra. London.

Allen, M. 1981. *Marsilio Ficino and the Phaedran Charioteer*. Berkeley/Los Angeles.

Aquilecchia, G. 1991. *Le opere italiane di Giordano Bruno*. Naples.

——. (1993a). *Schede bruniane (1950–1991)*. Rome.

——. 1993b. "Tre schede su Bruno e Oxford." *Giornale critico della filosofia italiana* 72.

——. 1995. "I *Massimi Sistemi di Galileo e la Cena di Bruno*. Paper delivered at the Accademia dei Lincei, Rome, April 22, 1995, and published in *Letture Galileiane* 3, (Florence): 485–508.

——. 1996. "Appunti su Bruno e le traduzioni." In *Giordano Bruno: Testi e traduzioni*, ed. H. Gatti. Rome.

——. 1997. "Giordano Bruno at Oxford." In *Giordano Bruno, 1583–1585: The English Experience*, ed. M. Ciliberto and N. Mann. Florence.

Aristotle. 1984. *The Complete Works*, 2 vols., ed. J. Barnes. Princeton.

Atanassievitch, X. 1923/1972. *The Metaphysical and Geometrical Doctrine of Bruno as Given in His Work "De Triplici Minimo."* St. Louis. Mo.

Bacon, F. 1858–68. *The Works*. Ed. J. Spedding, R. L. Ellis, and D. D. Heath. 15 vols. London/New York.

Badaloni, N. 1988. *Giordano Bruno tra cosmologia ed etica*. Bari.

——. 1991. "L'impulso del negativo: La *vanitas* in Giordano Bruno." In *Filosofia e cultura: per Eugenio Garin*. Rome.

——. 1994. "Riflessioni sul tema dell' 'individuo' nella concezione metafisica e morale di G. Bruno." *Nouvelles de la république des lettres* 2.

——. 1997. "Sulla struttura del tempo in Giordano Bruno." *Bruniana e campanelliana*, anno III, 1:11–45.
Baldini, U. 1977. "Il corpuscolarismo italiano del seicento." In *Ricerche sull'atomismo del seicento*, ed. U. Baldini, G. Zanier, and P. Farina. Florence.
Barlowe, W. 1617. *A Briefe Discovery of the Idle Animadversions of Mark Ridley*. London.
——. 1618. *Magneticall Advertisements*. London.
Barnes, J. 1975. "Introduction." In *Aristotle's Posterior Analytics*. Oxford.
Barrow, J. D. 1992. *Pi in the Sky: Counting, Thinking, and Being*. Oxford.
Batho, G. 1960. "The Library of the 'Wizard Earl': Henry Percy, Ninth Earl of Northumberland." *Library* 15.
Beierwalters, W. 1973/1983. "Absolute Identity: Neoplatonic Implications in Schelling's *Bruno*." In *Contemporary German Philosophy*, vol. 2. University Park, Pa.
Bernart, L. de. 1986. *Immaginazione e scienza in Giordano Bruno*. Pisa.
——. 1994. "Bruno e i 'fondamenti' filosofici della teoria copernicana." *Nouvelles de la république des lettres* 2.
Bilinski, B. 1977. *Il pitagorismo di Niccolò Copernico*. Warsaw.
Blackwell, R. 1991. *Galileo, Bellarmine, and the Bible*. London.
Blum, P. R. 1990. "Giordano Bruno, Matthias Aquarius und die eklektische Scholastik." *Archiv für Geschichte der Philosophie* 72:275–300.
Blumenberg, H. 1975/1987. "Not a Martyr for Copernicanism: Giordano Bruno." In *The Genesis of the Copernican World*. Cambridge, Mass.
Blundeville, T. 1599. *The Art of Logike*. London.
——. 1602. *The Theoriques of the Seven Planets*. London.
Boas, M. 1951. "Bacon and Gilbert." *Journal of the History of Ideas* 12.
Boden, M. A. 1990. *The Creative Mind: Myths and Mechanisms*. London.
Bohm, D. 1980. *Wholeness and the Implicate Order*. London.
Bolzoni, L. 1995. *La stanza della memoria: Modelli letterari e iconografici nell'età della stampa*. Turin.
Bönker-Vallon, A. 1995. *Metaphysik und Mathematik bei Giordano Bruno*. Berlin.
——. 1996. "The Mathematics of Natural Philosophy in Giordano Bruno's Latin Work: An Example of Early Modern Thought." *Proceedings of the PMR Conference* 19/20:125–32.
——. 1997. "Meßtheoretische Grundlagen." *Bruniana e Campanelliana* 3, no. 1: 47–75.
Bossy, J. 1991. *Giordano Bruno and the Embassy Affair*. New Haven.
Bressii, M. 1581. *Metrices Astronomicae*. Paris.
Brickman, B. 1943. "Francesco Patrizi: On Space." *Journal of the History of Ideas* 4.
Bruno, G. 1879–91. *Opera latine conscripta*. Ed. F. Fiorentino, V. Imbriani, C. M. Tallarigo, F. Tocco, and V. Vitelli. 3 vols., 8 pts. Naples/Florence. (Facsimile reprint, Friedrich Fromman Verlag, 1962.)
——. 1955. *La cena de le ceneri*. Ed. G. Aquilecchia. Turin.
——. 1957. *Due dialoghi sconosciuti e due dialoghi noti*. Ed. G. Aquilecchia. Rome.
——. 1958. *Dialoghi italiani*. 2 vols. Florence.
——. 1964. *Praelectiones geometricae e Ars deformationum*. Ed. G. Aquilecchia. Rome.

——. 1965. *The Heroic Frenzies*. Trans. P. E. Memmo Jr. Chapel Hill, N.C.
——. 1977. *The Ash Wednesday Supper*. Ed. E. A. Gosselin and L. S. Lerner. Hamden, Conn. (Reprint Toronto, 1995.)
——. 1980. *Opere latine*. Ed. C. Monti. Turin.
——. 1991a. *De umbris idearum*. Ed. R. Sturlese. Florence.
——. 1991b. *On the Composition of Images, Signs, and Ideas*. Trans. C. Doria. Ed. D. Higgins. New York.
——. 1964/1992. *The Expulsion of the Triumphant Beast*. Ed. A. D. Imerti. Lincoln, Neb.
——. 1993. *Oeuvres complètes: Chandelier*. Paris.
——. 1994a. *Oeuvres complètes: Cabale du cheval pégaséen*. Paris.
——. 1994b. *Oeuvres complètes: Le souper des cendres*. Paris.
——. 1995. *Il sigillo dei sigilli: I diagrammi ermetici*. Ed. Ubaldo Nicola. Milan.
——. 1996. *L'arte della memoria: Le ombre delle idee*. Ed. Manuela Maddamma. Milan.
Burkert, W. 1972. *Lore and Science in Ancient Pythagoreanism*. Cambridge, Mass.
Burtt, E. A. 1925. *The Metaphysical Foundations of Modern Physical Science*. London.
Buteonis, I. 1559. *De quadratura circuli libri duo*. Lyons.
Butterfield, H. 1950. *The Origins of Modern Science*. London.
Cabeo, N. 1629. *Philosophia magnetica*. Ferrara.
Cambi, M. 1993. "Il *De magia* e il recupero della sapienza originaria: Scrittura e voce nelle strategie magiche di Giordano Bruno." *Archivio di storia della cultura* 4.
Canone, E., ed. 1992. *Giordano Bruno: Gli anni napoletani e la "peregrinatio" europea*. Cassino.
Cantabrigiense, G. P. 1584. *Ad Alexandrum Dicsonum admonituncula de artificiosae memoriae quam publicè profitetur vanitate*. London.
Capek, M. 1987. "The Conflict between the Absolutist and the Relationist Theory of Time before Newton." *Journal of the History of Ideas* 48:603–6.
Casini, P. 1994. "Il mito pitagorico e la rivoluzione astronomica." *Rivista di filosofia* 85, no. 1.
Cassirer, E. 1927/1963. *The Individual and the Cosmos in Renaissance Philosophy*. New York.
Chappell, V. 1986. "The Theory of Ideas." In *Essays on Descartes's Meditations*, Ed. R. Rorty. Berkeley/Los Angeles.
Chevalley, C. 1984. "Kepler et Galilée dans la bataille des *Siderius Nuncius* (1610–1611)." In *Novità celesti e crisi del sapere*, ed. P. Galluzzi. Florence.
Ciliberto, M., ed. 1979. *Lessico di Giordano Bruno*. Rome.
——. 1986. *La ruota del tempo: Interpretazione di Giordano Bruno*. Rome.
——. 1990. *Giordano Bruno*. Rome/Bari.
——. 1994. "Tre note bruniane." *Rivista di storia della filosofia* 2.
Citolini, A. 1561. *La tipocosmia*. Venice.
Clagett, M. 1961. *The Science of Mechanics in the Middle Ages*. Madison, Wis.
Clarke, D. M. 1982. *Descartes' Philosophy of Science*. Manchester.
Clucas, S. 1997. "Walter Warner's Psychological Theories and Brunian Mnemonics." In *Giordano Bruno, 1583–1585: The English Experience*, ed. M. Ciliberto and N. Mann. Florence.

Clulee, N. H. 1988. *John Dee's Natural Philosophy: Between Science and Religion*. London.
Cohen. I. B. 1985. *Revolution in Science*. Cambridge, Mass.
Copenhaver, B. P. 1990. "Natural Magic, Hermeticism, and Occultism in Early Modern Science." In *Reappraisals of the Scientific Revolution*. ed. D. C. Lindberg and R. S. Westman. Cambridge/New York.
——. ed. 1992. *Hermetica*. Cambridge/New York.
Copenhaver, B. P., and C. B. Schmitt. 1992. *Renaissance Philosophy*. Oxford/New York.
Copernicus, N. 1543. *De revolutionibus orbium caelestium*. Norimberg.
——. 1543/1978. *On the Revolutions*. Ed. J. Dobrzycki, Trans. E. Rosen. Baltimore.
——. *Three Copernican Treatises*. Trans. and ed. E. Rosen. New York.
Cottingham, J. 1993. "A New Start? Cartesian Metaphysics and the Emergence of Modern Philosophy." In *The Rise of Modern Philosophy*, ed. T. Sorell. Oxford/New York.
Cox, V. 1992. *The Renaissance Dialogue*. Cambridge.
Dagron, T. 1994. "Giordano Bruno et la théorie des liens." *Etudes philosophiques*. 4.
Dee, J. 1564/1964. *Monas hieroglyphica*. Ed. and trans. C. H. Josten. London.
De Léon Jones, K. S. 1997. *Giordano Bruno and the Kabbalah*. New Haven.
Deregibus, A. 1980. *Bruno e Spinoza: La realtà dell'infinito e il problema della sua unità*. Vol.1, *Il concetto dell'infinito nel pensiero filosofico di Bruno*. Turin.
Descartes, R. 1961. *Regulae, or rules for the Direction of the Mind*. Trans. Laurence Lafleur. Indianapolis.
Dick, S. J. 1982. *Plurality of Worlds: The Origin of the Extraterrestrial Life Debate from Democritus to Kant*. Cambridge.
Dicson, A. 1583. *De umbra rationis et iudicii, sive de memoriae virtute prosopopeia*. London.
Duhem, P. 1913–59. *Le système du monde*, 10 vols. Paris.
Eamon, W. 1994. *Science and the Secrets of Nature: Books of Secrets in Medieval and Early Modern Culture*. Princeton.
Eddington, A. S. 1920. *Space, Time and Gravitation*. Cambridge.
Euclid of Megara. 1570. *The Elements of Geometrie*. Ed. H. Billingsley and J. Dee. London.
Farley-Hills, D. 1992. "The 'Argomento' of Bruno's *De gli eroici furori* and Sidney's *Astrophil and Stella*." *Modern Language Review* 87, no. 1: 1–17.
Feingold, M. 1983. *The Mathematicians' Apprenticeship: Science, Universities and Society in England, 1560–1640*. Cambridge.
Feyerabend, P. K. 1975. *Against Method*. London.
Ficino, M. 1956. *Théologie platonicienne de l'immortalité des ames*. Ed. Marcel. Paris
——. 1975–. *The Letters of Marsilio Ficino: Translated from the Latin by Members of the Language Department of the School of Economic Science*. London.
——. 1989. *Three Books on Life*. Binghamton, N.Y.
Field, J. 1988. *Kepler's Geometrical Astronomy*. London.
——. 1997. *The Invention of Infinity*. Oxford.
Firpo, L. 1993. *Il processo di Giordano Bruno*. Rome.
Freudenthal, G. 1983. "Theory of Matter and Cosmology in William Gilbert's *De magnete*." *Isis* 74.

Furley, D. 1981. "The Greek Theory of the Infinite Universe." *Journal of the History of Ideas* 42, no. 4.
Gabbey, Alan. 1990. "Mechanics: One Revolution or Many?" In *Reappraisals of the Scientific Revolution*, ed. D. C. Lindberg and R. S. Westman. Cambridge.
Galilei, G. 1890–1909. *Opere*. Ed. Favaro. Florence/Naples.
Galluzzi, P., ed. 1984. *Novità celesti e crisi del sapere*. Florence.
Garber, D. 1986. "*Semel in vita*: The Scientific Background to Descartes' *Meditations*." In *Essays on Descartes' Meditations*, ed. A. O. Rorty. Berkeley/Los Angeles.
Garin, E. 1954. *Medioevo e rinascimento*. Bari.
——. 1961. *La cultura filosofica del rinascimento italiano*. Florence.
——. 1975. *Rinascite e rivoluzioni: Movimenti culturali dal XIV al XVIII secolo*. Bari/Rome.
Garin, E., comp. 1955. *Testi umanistici sull'ermetismo*. Rome.
Gatti, H. 1989. *The Renaissance Drama of Knowledge: Giordano Bruno in England*. London.
——. 1993. *The Natural Philosophy of Thomas Harriot*. Oxford.
——. 1994a. "Smitho: Un mediocre o un saggio?" *Nouvelles de la république des lettres* 2.
——. 1994b. "*Telesio, Giordano Bruno e Thomas Harriot*." In *Accademia cosentina: Atti 1991–92*. Cosenza.
——. 1995. "Luigi Firpo: *Il processo di Giordano Bruno*." *British Journal for the History of Philosophy* 3, no, 1.
——. 1996a. "Coleridge's Reading of Giordano Bruno." *Wordsworth Circle*, Summer.
——. 1996b. "L'idea di riforma nei dialoghi italiani di Giordano Bruno." *Nouvelles de la république des lettres* 2.
——, ed. 1996. *Giordano Bruno: Testi e traduzioni*. Rome.
——. 1997. "Giordano Bruno's *Ash Wednesday Supper* and Galileo's *Dialogue of the Two Major World Systems*." *Bruniana e Campanelliana*, anno III, 2.
Gaukroger, S., and R. Ariew, eds. 1991. *The Uses of Antiquity: The Scientific Revolution and the Classical Tradition*. Dordrecht/Boston/London.
Gentile, G. 1991. *Giordano Bruno e il pensiero del rinascimento*. Ed. E. Garin. Florence.
Giaquinto, M. 1992. "Visualizing as a Means of Geometrical Discovery." *Mind and Language* 7, no. 4.
——. 1993. "Visualizing in Arithmetic." *Philosophy and Phenomenological Research* 53.
——. 1994. "Epistemology of Visual Thinking in Elementary Real Analysis." *British Journal for the Philosophy of Science*, June.
Gilbert, W. 1600/1958. *De magnete*. Ed. P. F. Mottelay. London.
——. 1651. *De mundo nostro sublunari philosophia nova*. Amsterdam.
Giusso, L. 1955. *Scienza e filosofia in Giordano Bruno*. Naples.
Gobry, I. 1992. *Pythagore*. Paris.
Goldstein, B. R., and P. Barker. 1995. "The Role of Rothmann in the Dissolution of the Celestial Spheres." *British Journal of the History of Science* 28:385–403.
Gosselin, E. A. 1988. "Bruno's French Connection: A Historiographical Debate." In *Hermeticism and the Renaissance: Intellectual History and the Occult in Early Modern Europe*, ed. I. Merkel and A. G. Debus. Washington, D.C.

Gottschalk, H. B. 1980. *Heraclitus of Pontus*. Oxford.
Grafton, A. 1983. "Protestant versus Prophet: Isaac Casaubon on Hermes Trismegistus." *Journal of the Warburg and Courtauld Institutes* 46.
Granada, M. 1990. "L'interpretazione bruniana di Copernico e la *Narratio prima* di Rheticus." *Rinascimento* 30.
——. 1992. "Bruno, Digges, Palingenio: Omogeneità ed Eterogeneità nella concezione dell'universo infinito." *Rivista di storia della filosofia* 47, no. 1.
——. 1994. "Il rifiuto della distinzione fra *potentia absoluta* e *potentia ordinata* di Dio e l'affermazione dell'universo infinito in Giordano Bruno." *Storia della filosofia* 3.
——. 1996. *El debate cosmológico en 1588: Bruno, Brahe, Rothmann, Ursus, Röslin*. Naples.
——. 1997. "Thomas Digges, Giordano Bruno e il copernicanesimo in Inghilterra." In *Giordano Bruno, 1583–1585: The English Experience*, ed. M. Ciliberto and N. Mann. Florence.
Grant, E. 1994. *Planets, Stars, and Orbs: The Medieval Cosmos, 1200–1687*. Cambridge.
Greenburg, S. T. 1950. *The Infinite in Giordano Bruno, with a translation of His Dialogue "Concerning the Cause, Principle, and One."* New York.
Guthrie, K. S., comp. 1987. *The Pythagorean Sourcebook and Library*. Ed. D. R. Fideler. Grand Rapids, Mich.
Harvey, G. 1913. *Marginalia*. Ed. C. G. Moore Smith. Oxford.
Henry, J. H. 1979. "Francesco Patrizi da Cherso's Concept of Space and Its Later Influence." *Annals of Science* 36:549–75.
Hiley, B. J., and F. D. Peat, eds. 1989. *Quantum Implications: Essays in Honour of David Bohm*. London/New York.
Hofstadter, D. R. 1979. *Gödel, Escher, Bach: An Eternal Golden Braid*. New York.
——. 1985. *Metamagical Themes: Questing for the Essence of Mind and Pattern*. New York.
Hofstadter, D. R., and D. C. Dennett, eds. 1981. *The Mind's I*. New York.
Horowitz, I. R. 1952. *The Renaissance Philosophy of Giordano Bruno*. New York.
Huffman, C. A. 1993. *Philolaus of Croton: Pythagorean and Presocratic*. Cambridge.
Ingegno, A. 1978. *Cosmologia e filosofia nel pensiero di Giordano Bruno*. Florence.
——. 1985. *La sommersa nave della religione: Studio sulla polemica anticristiana del Bruno*. Naples.
——. 1987. *Regia pazzia: Bruno lettore di Calvino*. Urbino.
Jacquot, J. 1952. "Thomas Harriot's Reputation for Impiety." *Notes and Records of the Royal Society* 9:164–87.
——. 1974. "Harriot, Hill, Warner and the New Philosophy." In *Thomas Harriot: Renaissance Scientist*. Oxford.
Jardine, N. 1984. *The Birth of History and Philosophy of Science*. Cambridge.
Jolley, N. 1990. *The Light of the Soul: Theories of Ideas in Leibniz, Malebranche, and Descartes*. Oxford.
Kargon, R. 1966. *Atomism in England from Harriot to Newton*. Oxford.
Kelly, S. 1965. *The "De mundo" of William Gilbert*. Amsterdam.
Kempe, A. 1995. *Concepts of Space in Greek Thought*. Leiden.

Kepler, J. 1937–75. *Gesammelte Werke*. 17 vols. Munich.
——. 1997. *The Harmony of the World*. Ed. and trans. E. J. Aiton et al. Philadelphia.
Kingsley, Peter. 1994. "From Pythagoras to the *Turba philosophorum*: Egypt and Pythagorean Tradition." *Journal of the Warburg and Courtauld Institutes* 57.
Knorr, W. R. 1986. *The Ancient Tradition of Geometric Problems*. New York.
Koyré, A. 1957. *From the Closed World to the Infinite Universe*. Baltimore.
——. 1966. *Etudes galiléennes*. Paris.
Kraye, J. 1993. "The Philosophy of the Italian Renaissance." In *Routledge History of Philosophy*, vol. 4, *The Renaissance and Seventeenth-Century Rationalism*, ed. G. H. R. Parkinson. London.
Kristeller, P. O. 1964. "Giordano Bruno." In *Eight Philosophers of the Italian Renaissance*. Stanford.
Kuhn, T. S. 1957. *The Copernican Revolution: Planetary Astronomy in the Development of Western Thought*. Cambridge, Mass.
——. 1962. *The Structure of Scientific Revolutions*. Chicago.
Lindberg, D. C., and R. S. Westman, eds. 1990. *Reappraisals of the Scientific Revolution*. Cambridge.
Lucretius. 1910. *De rerum natura*. Trans. and ed. C. Bailey. Oxford.
Mair, J. 1938. *Le traité "De l'infini."* Ed. H. Elie. Paris.
Mann, P. A. 1970. *The Infinite Worlds of Giordano Bruno*. Springfield, Ill.
Marenco, F. 1968. *Arcadia Puritana: l'uso della tradizione nella prima "Arcadia" di Sir Philip Sidney*. Bari.
Maspero, R. 1989. "Scienza e copernicanesimo in Bruno: Principali orientamenti della critica dal 1950 ad oggi." *Rivista di storia della filosofia* 2, no. 1.
Massa, D. 1973. "Giordano Bruno and the Top-Sail Experiment." *Annals of Science* 30.
——. 1977. "Giordano Bruno's Ideas in Seventeenth-Century England." *Journal of the History of Ideas* 38.
Mates, B. 1989. *The Philosophy of Leibniz: Metaphysics and Language*. Oxford.
McColley, G. 1937. "William Gilbert and the English Reputation of Giordano Bruno." *Annals of Science* 2.
McGuire, J. E., and P. M. Rattansi. 1966. "Newton and the Pipes of Pan." *Notes and Records of the Royal Society* 21.
McMullin, E. 1987. "Bruno and Copernicus." *Isis* 78.
——. 1990. "Conceptions of Science in the Scientific Revolution." In *Reappraisals of the Scientific Revolution*, ed. D. C. Lindberg and R. S. Westman. Cambridge.
Mendoza, R. G. 1995. *The Acentric Labyrinth: Giordano Bruno's Prelude to Contemporary Cosmology*. Shaftesbury/Rockport/Brisbane.
Merkel, I., and A. G. Debus, eds. 1988. *Hermeticism and the Renaissance: Intellectual History and the Occult in Early Modern Europe*. Washington, D.C.
Michel, P. H. 1957. "L'atomisme de Giordano Bruno." In *La science au seizième siècle*. Paris.
——. 1964. "Les notions de continu et de discontinu dans les systèmes physiques de Bruno et de Galilée." In *Mélanges Alexandre Koyré*, vol. 2. Paris.
——. 1962/1973. *The Cosmology of Giordano Bruno*. Ed. R. E. W. Madison. Ithaca, N.Y.
Miele, M. 1992. "L'organizzazione degli studi domenicani di Napoli al tempo di

Giordano Bruno." In *Giordano Bruno: Gli anni napoletani e la "peregrinatio" europea*, ed. E. Canone. Cassino.
Miller, A. I. 1984. *Imagery in Scientific Thought: Creating Twentieth-Century Physics*. Boston.
Montaigne, M. de. 1595/1958. *Essays*. Ed. J. M. Cohen. Harmondsworth.
Monti, C. 1994. "Incidenza e significato della tradizione materialista antica nei poemi latini di Giordano Bruno: La mediazione di Lucrezio." *Nouvelles de la république des lettres* 2.
Morowitz, H. J. 1981. "Rediscovering the Mind." In *The Mind's I*, ed. D. R. Hofstadter and D. C. Dennett. New York.
Nizolio, M. 1953. *De principiis*. In *Archivio di filosofia*. Rome.
Nowicki, A. 1967. "Un autografo inedito di Giordano Bruno in Polonia." *Atti dell'Accademia di scienze morali e politiche* 78.
O'Meara, D. J. 1991. *Pythagoras Revived*. Oxford.
Ordine, N. 1987/1996. *Giordano Bruno and the Philosophy of the Ass*. New Haven.
Osborn, H. F. 1894. *From the Greeks to Darwin: An Outline of the Development of the Evolution Idea*. New York.
Pacioli, L. 1509. *Divina proportione*. Venice.
Pagel, W. 1951. "Giordano Bruno: The Philosophy of Circles and the Circular Movement of the Blood." *Journal of the History of Medicine and Allied Sciences* 6.
Palingenius, M. 1531. *Zodiacus vitae*. Venice.
——. 1565/1947. *The Zodiake of Life*. Trans. B. Googe. Ed. R. Tuve. New York.
Parkinson, G. H. R., ed. 1993. *Routledge History of Philosophy. Vol. 4, The Renaissance and Seventeenth-Century Rationalism*. London.
Pena, J. 1557. *Euclidis optica et catoptica*. Paris.
Penrose, R. 1989. *The Emperor's New Mind*. Oxford.
Pérez-Ramos, A. 1993. "Francis Bacon and Man's Two-Faced Kingdom." In *Routledge History of Philosophy*, Vol. 4, *The Renaissance and Seventeenth-Century Rationalism*, ed. G. H. R. Parkinson. London.
Perkins, G. 1584a. *Antidicsonus*. London.
——. 1584b. *Libellus de memoria: Ad Alexandrum Dicsonum admonituncula de artificiosae memoriae quam publicè profitetur vanitate*. London.
Perugini, R. 1984. *La memoria creativa: Architettura e arte tra Rinascimento e Illuminismo*. Rome.
Plato. 1961. *The Collected Dialogues*. Ed. E. Hamilton and H. Cairns. Princeton.
Popkin, R. 1960/1979. *History of Skepticism from Erasmus to Spinoza*. Berkeley/Los Angeles.
Popper, K. 1959. *The Logic of Scientific Discovery*. London.
Post, H. R. 1971. "Correspondence, Invariance and Heuristics." *Studies in the History and Philosophy of Science* 2. Reprint 1993. *Correspondence, Invariance and Heuristics: Essays in Honour of Heinz Post*. Ed. S. French and H. Kamminga. Boston.
Prins, J. (N.d.) *Walter Warner on Space and Time*. Durham Thomas Harriot Seminar Occasional Paper no. 16.
Prior, M. E. 1968. "Bacon's Man of Science." In *Essential Articles for the Study of Francis Bacon*, ed. B. Vickers. London.

Ptolemy. 1984. *The Almagest*. Ed. and trans. G. Toomer. London.
Pumphrey, S. "William Gilbert's Magnetic Philosophy." Ph.D diss., Warburg Institute. London.
Putnam, H. 1975. *Mind, Language and Reality*. Cambridge.
Rattansi, P. M. 1996. "Reform as *Renovatio* in Sixteenth-Century Science and Medicine." *Nouvelles de la république des lettres* 2:97–115.
Regiomontanus. 1533/1967. *On Triangles*. Ed. B. Hughes. Madison, Wis.
Reuchlin, G. 1517. *De arte cabalistica libri tres*. Basel.
Ricci, S. 1990. *La fortuna del pensiero di Giordano Bruno, 1600–1750*. Florence.
——. 1991. "La recezione del pensiero di Giordano Bruno in Francia e Germania: Da Diderot a Schelling." In *Giornale critico della filosofia italiana* 11.
——. 1994. "Scienza e vita civile da Giordano Bruno ai Lincei." In *Una filosofica milizia: Tre studi sull'Accademia dei Lincei*. Udine.
Ridley, M. 1613. *A Short Treatise of Magnetical Bodies and Motions*. London.
——. 1617. *Magneticall Animadversions*. London.
Rorty, R. 1980. *Philosophy and the Mirror of Nature*. Oxford.
——. 1991. "Science as Solidarity." in *Objectivity, Relativism and Truth*. Cambridge.
Rose, P. L. 1975. *The Italian Renaissance of Mathematics: Studies on Humanists and Mathematicians from Petrarch to Galileo*. Geneva.
Rossi, P. 1953. "Rettorica e antimetafisica nel *De principiis* di M. Nizolio." In *La crisi dell'uso dogmatico della ragione*. Rome/Milan.
——. 1974. *Francesco Bacone: Dalla magia alla scienza*. Turin.
——. 1984. "Ants, Spiders, Epistemologists." In *Lessico intellettuale europeo: Bacon*. Rome.
Salvestrini, V., and L. Firpo, eds. 1958. *Bibliografia di Giordano Bruno, 1582–1950*. Florence.
Sanches, F. 1581/1988. *That Nothing Is Known*. Ed. E. Limbrick and D. Thomson. Cambridge.
Sarpi, P. 1619/1974. *Istoria del Concilio tridentino*. Ed. C. Vivanti. 2 vols. Turin.
Sarton, G. 1957. *Six Wings: Men of Science in the Renaissance*. Bloomington, Ind.
Schmitt, C. 1983. *Aristotle in the Renaissance*. Cambridge, Mass.
Schmitt, C., Q. Skinner, and E. Kessler, eds. 1988. *The Cambridge History of Renaissance Philosophy*. Cambridge.
Shanker, S. 1988. *Gödel's Theorem in Focus*. London.
Shirley, J. 1951. "Binary Numeration before Leibniz." *American Journal of Physics* 19:452–54.
——. ed. 1974. *Thomas Harriot: Renaissance Scientist*. Oxford.
——. 1983. *Thomas Harriot: A Biography*. Oxford.
Shumaker, Wayne. 1989. *Natural Magic and Modern Science: Four Treatises (1590–1657)*. New York.
Sidney, P. 1593/1977. *The Countess of Pembroke's Arcadia*. Harmondsworth.
——. 1595/1973. *An Apology for Poetry*. Manchester.
Singer, D. 1950. *Giordano Bruno: His Life and Thought*. New York.
Sorabji, R. 1983. *Time, Creation and the Continuum: Theories in Antiquity and the Early Middle Ages*. Ithaca, N.Y.
——. 1988. *Matter, Space, and Motion: Theories in Antiquity and Their Sequel*. Ithaca, N.Y.

Sorell, T., ed. 1993. *The Rise of Modern Philosophy*. Oxford.
Spampanato, V. 1921/1988. *Vita di Giordano Bruno con documenti editi e inediti*. Ed. N. Ordine. Rome.
Sprat, T. 1667/1959. *The History of the Royal Society of London*. Ed. J. I. Cope and H. W. Jones. St. Louis/London.
Spruit, L. 1988. *Il problema della conoscenza in Giordano Bruno*. Naples.
——. 1995. "Giordano Bruno." In *Species Intelligibilis: From Perception to Knowledge*. Leiden.
Sturlese, R. 1987. "Un nuovo autografo del Bruno con una postilla sul *De umbra rationis* di A. Dicson." *Rinascimento* 27.
——, ed. 1987. *Bibliografia, censimento e storia delle antiche stampe di Giordano Bruno*. Florence.
——. 1990. "Il *De imaginum, signorum et idearum compositione* di Giordano Bruno ed il significato filosofico dell'arte della memoria." *Giornale critico della filosofia italiana*, May–August.
——. 1992a. "'Averroe quantumque arabo et ignorante di lingua greca . . .': Note sull'avveroismo di Giordano Bruno." *Giornale critico della filosofia italiana* 71.
——. 1992b. "Per un'interpretazione del *De umbris idearum* di Giordano Bruno." *Annali della Scuola Normale Superiore di Pisa*, 3d ser., 22, no. 3.
——. 1994. "Le fonti del *Sigillus sigillorum* del Bruno, ossia il confronto con Ficino a Oxford sull'anima umana." *Nouvelles de la république des lettres* 2.
Sturlese-Pagnoni, M. R. 1985. "Su Bruno e Tycho Brahe." *Rinascimento* 25.
——. 1986. "Postille autografe di John Toland allo Spaccio del Bruno." *Giornale critico della filosofia italiana* 65.
Swerdlow, N. M. 1975. "On Copernicus' Theory of Precession." In *The Copernican Achievement*, ed. R. Westman. Los Angeles.
Swerdlow, N. M., and O. Neugebauer. 1984. *Mathematical Astronomy in Copernicus's "De revolutionibus."* 2 vols. New York.
Telesius, B. 1570. *De rerum natura*. Naples.
——. 1586. *De natura iuxta propria principia*. Naples.
Teti, D. 1988. *Alcmeone e Pitagora*. Reggio Calabria.
Thom, J. C. 1995. *The Pythagorean Golden Verses*. Leiden/New York.
Thorndike, L. 1923–58. *History of Magic and Experimental Science*. 8 vols. New York.
Tocco, F. 1889. *Le opere latine di G. Bruno esposte e confrontate con le italiane*. Florence.
Toynbee, P. J. 1894. "Dante's References to Pythagoras." In *Index of Proper Names in the Prose Works and Canzoniere of Dante*. Boston.
Trompf, G. W. 1991. "On Newtonian History." In *The Uses of Antiquity: The Scientific Revolution and the Classical Tradition*," ed. S. Gaukroger and R. Ariew. Dordrecht/Boston/London.
Tyard, P. de. 1578. *Deux discours de la nature du monde et de ses parties*. Paris.
Tymme T. 1612. *A Dialogue Philosophicall Wherein Natures secret closet is opened, and the cause of all motion in Nature shewed out of Matter and Form*. London.
Vasoli, C. 1988. *Filosofia e religione nella cultura del Rinascimento*. Naples.
——. 1991. "Francesco Patrizi sull'infinità dell'universo." In *Filosofia e cultura: per Eugenio Garin*. Rome.

Vedrine, H. 1967. *La conception de la nature chez Giordano Bruno*. Paris.
——. 1970. "L'influence de Nicolas de Cues sur Giordano Bruno." In *Niccolò Cusano agli inizi del mondo moderno*. Florence.
Vickers, B., ed. 1968. *Essential Articles for the Study of Francis Bacon*. London.
——. 1984. "Analogy versus Identity: The Rejection of Occult Symbolism." In *Occult and Scientific Mentalities in the Renaissance*. Cambridge.
——. ed. 1984. *Occult and Scientific Mentalities in the Renaissance*. Cambridge.
Walker, D. P. 1958. *Spiritual and Demonic Magic from Ficino to Campanella*. London.
Wallace, W. A. 1981. *Prelude to Galileo*. Dordrecht.
Waller, G. F. 1972. "'This Matching of Contraries': Bruno, Calvin and the Sidney Circle." *Neophilologus* 56: 331–43.
Weiner, A. D. 1980. "Expelling the Beast: Bruno's Adventures in England." *Modern Philology* 78: 1–13.
West, M. L. 1971. "The Cosmology of *De hebdomadis*." *Classical Quarterly* 21:365–88.
Westman, R. 1975a. "The Melanchthon Circle, Rheticus, and the Wittenberg Interpretation of the Copernican Theory." *Isis* 66.
——. ed. 1975b. *The Copernican Achievement*. Berkeley/Los Angeles.
——. 1977. "Magical Reform and Astronomical Reform: The Yates Thesis Reconsidered." In *Hermeticism and the Scientific Revolution*. Los Angeles.
——. 1984. "Nature, Art, and Psyche: Jung, Pauli, and the Kepler-Fludd Polemic." In *Occult and Scientific Mentalities in the Renaissance*, ed. B. Vickers. Cambridge.
——. 1990. "Proof, Poetics and Patronage: Copernicus's Preface to *De Revolutionibus*." In *Reappraisals of the Scientific Revolution*, ed. D. C. Lindberg and R. S. Westman. Cambridge.
Whewell, W. 1859. *History of the Inductive Sciences*. Cambridge.
Yates, F. 1936. *A Study of "Love's Labours Lost."* Cambridge.
——. 1947. *The French Academies of the Sixteenth Century*. London.
——. 1964. *Giordano Bruno and the Hermetic Tradition*. London.
——. 1966. *The Art of Memory*. Chicago.
——. 1982. *Lull and Bruno: Collected Essays*. Vol. 1. London/Boston.

Index